A Concise Gui
ANIMAL T
of Southern Africa

Louis Liebenberg

David Philip Publishers
Cape Town & Johannesburg

Credits: photographs by David Bristow facing pages 25 (top), 40 (top), 56, 72 (bottom) and 73 (bottom); all other photographs by Louis Liebenberg.

First published 1992 in southern Africa by David Philip Publishers (Pty) Ltd, 208 Werdmuller Centre, Claremont 7700, South Africa

ISBN 0-86486-230-X

Printed by Creda Press, Solan Road, Cape Town, South Africa

CONTENTS

ACKNOWLEDGEMENTS

I would like to thank:

My parents for their financial assistance and support, without which this book would not have been possible.

The late Dr. R. H. N. Smithers, Dr. N. J. Dippenaar, Dr. S. Endrödy-Younga, Dr. M. Mansel, Dr. A. Prins, Dr. G. McLachlan and Mrs C. Carr for advice and criticism.

The !Xõ trackers Bahbah, Jehjeh and Hewha of Ngwatle Pam and N!am!kabe, Kayate, N!ate and Boroh//xao of Lone Tree and Lefi Cooper of Lokgwabe in Botswana.

Mr T. Robson, Mr J. T. W. Nicholls, Mr J. Varty, Mr D. Varty, Mr T. Thompson and Mr Phillip Ndlovu of the Sabi Sand Nature Reserve.

The directors of the Natal Parks Board and the C.P.A. Nature Conservation Department for research permits.

The wardens and rangers of the following nature reserves where field research on spoor was conducted: Etosha Nature Reserve; Umfolozi Nature Reserve; St Lucia Nature Reserve; Giant's Castle Nature Reserve; Tsitsikamma Forest National Park; De Hoop Nature Reserve; Cape of Good Hope Nature Reserve; Table Mountain Nature Reserve; Cedarberg Wilderness Reserve; Kenneth Stainbank Nature Reserve, Durban; Naval Hill Reserve, Bloemfontein; and Krugersdorp Nature Reserve.

The owners and curators of the following zoological gardens where the spoor of animals in captivity have been studied: Bloemfontein Zoological Gardens; Johannesburg Zoological Gardens; National Zoological Gardens; Natal Zoological Gardens; Queens Park Zoological Gardens, East London; Hartebeespoortdam Snake and Animal Park; Tygerberg Zoological Gardens; Fitzsimons Snake Park, Durban; Centre for the Rehabilitation of Wildlife, Durban; The World of Birds, Cape Town; Africa Fauna Bird Park, Krugersdorp; Aromaland Zoological Gardens, Brackenfell; Strandfontein Snake, Crocodile and Reptile Park; private zoological collections of Mr G. Carpenter, Miss E. Jordaan and Mr P. and Mrs A. Krüger.

The directors and curators of the following museums where the feet of specimens have been studied: Transvaal Museum, Pretoria; South African Museum, Cape Town; National Museum, Bloemfontein; Natal Museum, Pietermaritzburg; Kaffrarian Museum, King William's Town.

Fotini Babaletakis for helping me draw the distribution maps and rearranging the spoor plates..

Mrs S. Louw, Mrs M. E. Clarke and Mrs S. Wyndham for typing various parts of the manuscript.

David and Marie Philip, Russell Martin, Ingrid Küpper and the other staff members of David Philip Publishers for the production of this book.

We would like to thank the following authors and publishers for permission to adapt or condense material from the books mentioned:

Roberts' Birds of Southern Africa (Cape Town: the John Voelcker Bird Book Fund, 1985);

Animal Life in Southern Africa (Cape Town: Nasou Limited, 1971) by D. J. Potgieter, P. C. du Plessis and S. H. Skaife;

A Field Guide to the Snakes of Southern Africa (London: Collins, 1970) by V. F. M. Fitzsimons;

The Mammals of the Southern African Subregion (Pretoria: University of Pretoria, 1983) by R. H. N. Smithers;

Field Guide to the Snakes and Other Reptiles of Southern Africa (Cape Town: Struik Publishers, 1988) by B. Branch;

African Insect Life (Cape Town: Struik Publishers, 1979) by S. H. Skaife, revised by John Ledger.

INTRODUCTION

This book was written specially for nature lovers, animal watchers, tourists, hikers and armchair naturalists who want a concise but accurate guide to help them identify the animal spoor they may encounter in southern Africa. Based on the author's authoritative and comprehensive *Field Guide to the Animal Tracks of Southern Africa* (David Philip, 1990), it provides detailed illustrations of the footprints of the more common species one is likely to come upon and sufficient information to assist one in making the right identification. In particular, the book is intended as a practical guide for younger enthusiasts, seeking to enhance their environmental awareness by revealing the multitude of tracks and signs in the wild.

Even keen nature lovers are often unaware of the wealth of animal life around them, simply because most animals are rarely seen. I once encountered a group of about a dozen hikers who walked right over a perfectly clear Leopard spoor. Not one of them noticed it, simply because they were not 'spoor conscious'. To them the Leopard simply did not exist. Yet to find a fresh Leopard spoor in the wilderness adds an exciting new dimension to hiking in the wilds. The Leopard is a stealthy animal that is rarely seen. But its spoor tells you that it is there.

To the nature lover, spoor may reveal the activities of many animals that would otherwise never be seen. To the untrained eye the wilderness can appear desolate, but to someone who is at least 'spoor conscious' it will be full of signs of wildlife. Even if you never see the animals, the knowledge that they are there is enough. By reconstructing their movements from their footprints, you may be able to visualise the animals and in your imagination actually 'see' them. In this way a whole story may unfold – a story of what happened when no one was looking.

ABBREVIATIONS

HB Length of head and body
Ht Height
T Tail length
TL Total length

The system of numbering the spoor of particular animals that is followed in this book corresponds to that originally adopted in the author's comprehensive *Field Guide to the Animal Tracks of Southern Africa* (David Philip, 1990), to which readers are referred for fuller information and a greater variety of spoor.

SPOOR IDENTIFICATION

The art of tracking involves each and every sign of animal presence that can be found in nature, including ground spoor, vegetation spoor, scent, feeding signs, urine, faeces, saliva, pellets, territorial signs, paths and shelters, vocal and other auditory signs, visual signs, incidental signs, circumstantial signs and skeletal signs. In this book only footprints are dealt with, as they are the easiest to identify. Footprints provide the most detailed information on the identity, movements and activities of animals, and once a trail has been identified, other signs can be studied in more detail. Footprints therefore offer a valuable introduction to the art of tracking, a science that might otherwise prove inaccessible to the inexperienced naturalist.

The spoor illustrations in this book, which are exact studies made under ideal conditions, may be regarded as generalised models that have been used to simplify spoor interpretation. In reality one will probably never find two animals with exactly identical footprints. One therefore needs abstractions to identify characteristic features of the spoor of different species.

A further advantage of using models is that it gives one a preconceived image that improves the chances of recognising spoor which may otherwise be overlooked. Preconceived images play an important role in the recognition of patterns in nature. However, with a preconceived image in mind, one tends to 'recognise' patterns in markings that may have been made by other animals, or even random markings. One must be careful not to be prejudiced and see what one wants to see.

While species may be recognised by some general characteristics, each individual animal's spoor differs in very subtle ways, and it is in principle possible to identify an individual animal from its spoor. So, for example, Kalahari trackers can identify the antelope they have shot from the rest of the herd and will track down that individual animal. Apart from the functional and environmental adaptations of the species, an individual animal's spoor may vary according to its age, mass, sex, condition, and the terrain as well as random variations. It may also have a unique way of walking or a peculiar habit that distinguishes it from other individuals.

While the spoor of most of the larger mammals and birds can be identified as belonging to a particular species, the spoor of the smaller animals may only be identified as belonging to a genus, family or order. The smaller the animal, the more difficult it becomes to distinguish its spoor from that of similar species, and while some mammal families may consist of only a few species, insect families may contain thousands. Some of the antelope species may have spoor characteristics typical of the species, but variations may occur that are similar to those of other species.

The best footprints are usually found in damp, slightly muddy earth, in wet sand, in a thin layer of loose dust on firm substrate, or in snow. Ideal wet conditions are found along streams, rivers, waterholes, dams, vleis, beaches, after rain or in the morning when the sand is still damp from the dew. Puddles that have just dried out, leaving a thin layer of mud over a firm substrate, are ideal for tracks of small animals. Dirt roads and paths may have a thin layer of very fine dust on firm ground that can reveal the finest detail of the spoor. Usually, however, footprints are partially obliterated, and one should walk up and down the trail to find the best imprints. Even if no clear footprints can be found, one can collect bits of information by studying several footprints and piece them together to compile an image of the complete spoor.

When studying spoor in loose sand, one should try to visualise the shape of the footprint before the loose sand grains slid together to obliterate the well-defined features. As much information may be lost in loose sand, it is not always possible to distinguish the spoor of similar species, so it would have to be considered as belonging to any one of several possibilities, until further evidence is gathered.

It should be kept in mind that footprints may be distorted owing to slipping and twisting of the feet on the ground. When the animal is walking on a slope or running, the feet may slip, so the spoor will appear elongated or warped. If the fore and hind spoor are superimposed, it may look like an elongated spoor, or the toes of the fore footprint may be confused with those of the hind. When trotting or running, the animal's mass is supported mainly on the toes and only part of the intermediate pads may show, or, in the case of mongooses, the proximal pads may not show at all. On hard ground padded toes may not show and only claw marks may be seen.

If the spoor could be that of several possible species, the distribution maps in this field guide should be consulted to eliminate those that do not occur in that locality. Habitat and habits, such as sociability and daily rhythm, as well as feeding signs and faeces, should also be considered to narrow down the range of possibilities.

An easy way to compare similar spoor is to trace the outline of the spoor illustrations on transparent tracing paper and to place it over similar spoor to see how they differ. In the process of tracing you will get to know the spoor's distinctive features, which will help you recognise them in the field.

Note that spoor illustrations printed natural size, or close to natural size, appear to be larger than the actual spoor on the ground. This is due to an optical illusion created by the greater contrast between the black ink and white paper, compared to the more subtle shadows in the actual spoor on the ground. This discrepancy should also be kept in mind when drawing spoor, since one tends to draw it smaller than the actual size to make it appear the same size. It is therefore essential to measure each detail of the spoor.

Spoor identification requires not only a great deal of knowledge, but also skill and experience. Although the inexperienced naturalist should in principle be able to use this book to identify near-perfect spoor in ideal conditions, the accurate identification of imperfect spoor, especially in loose sand, may only be possible after considerable experience.

INVERTEBRATES

1 Earthworm

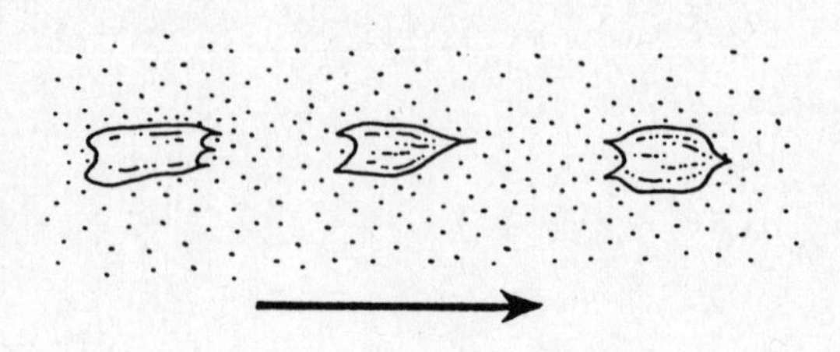

2.1 Snail

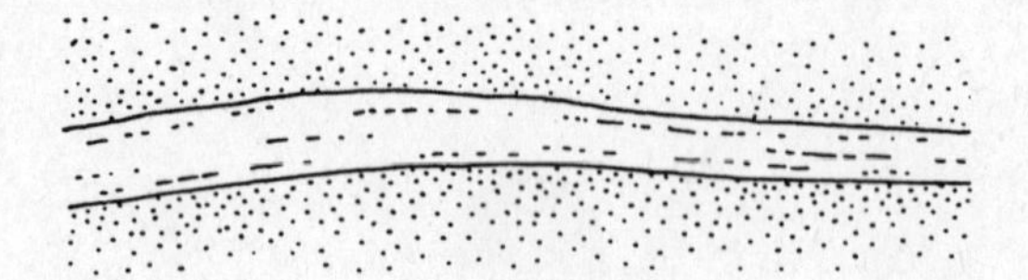

2.2 Slug

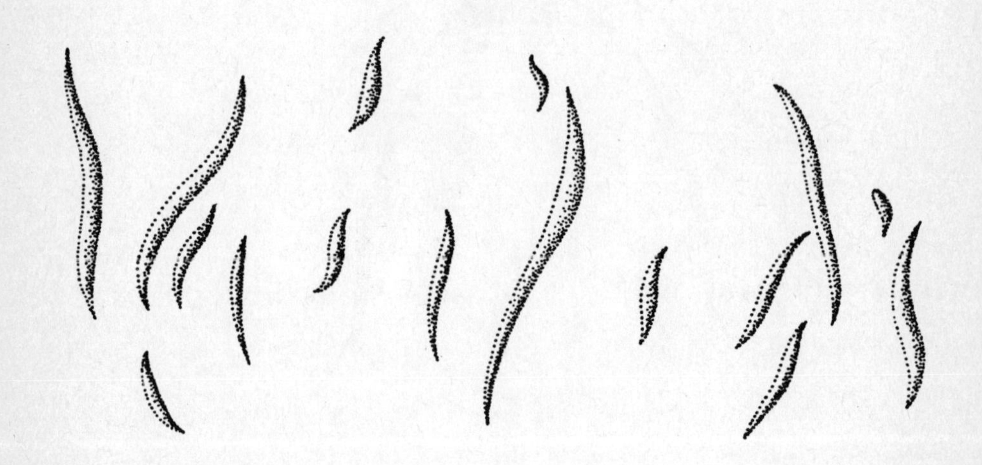

Leaf rolling in wind

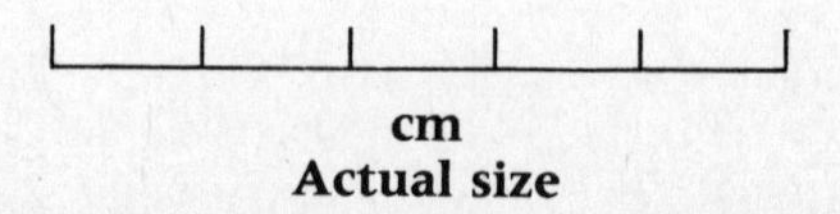

Actual size

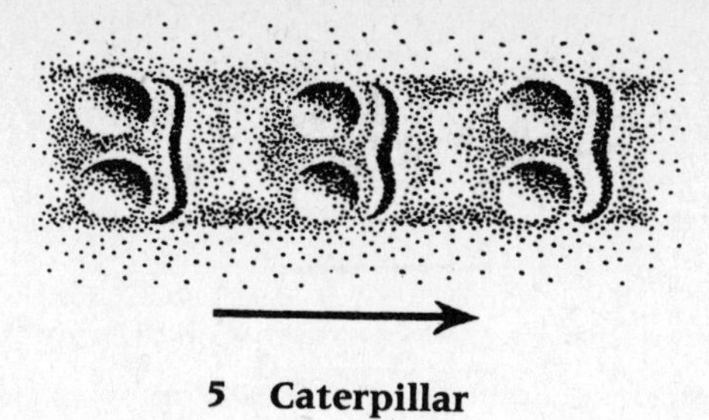

5 Caterpillar

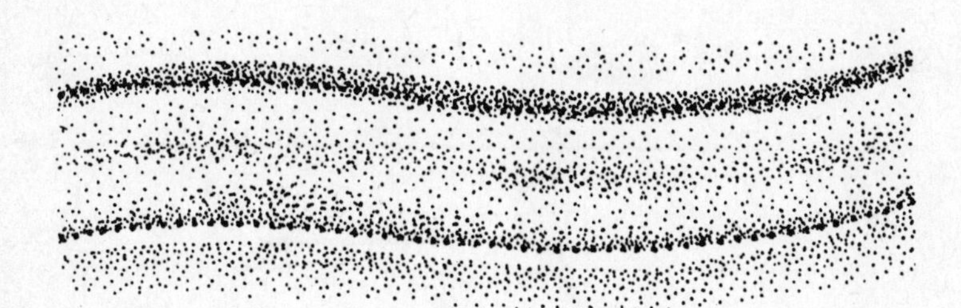

3 Millipede

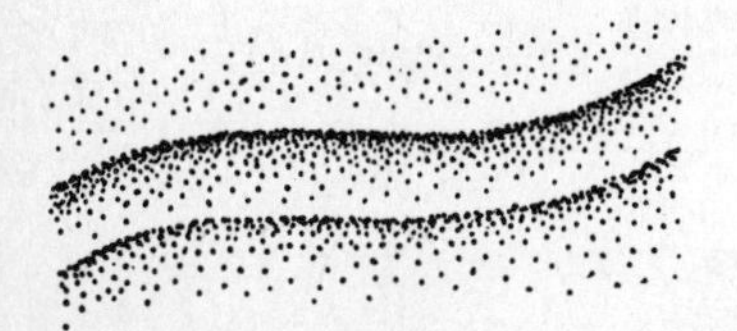

4a Antlion larva spoor

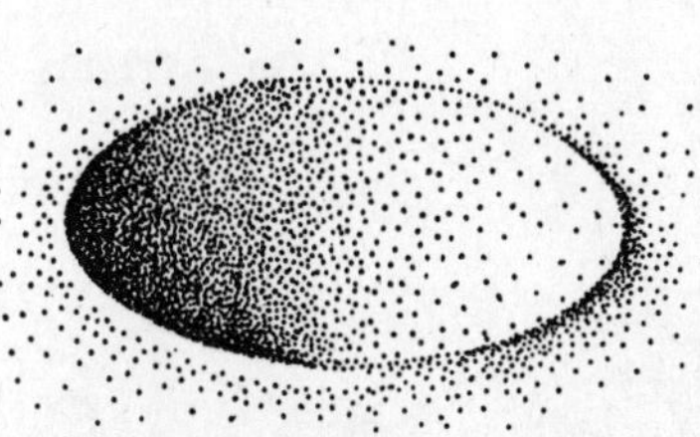

4b Antlion larva pit

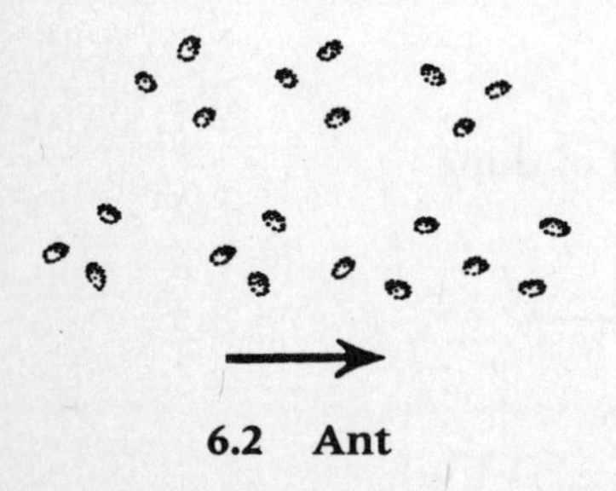

6.2 Ant

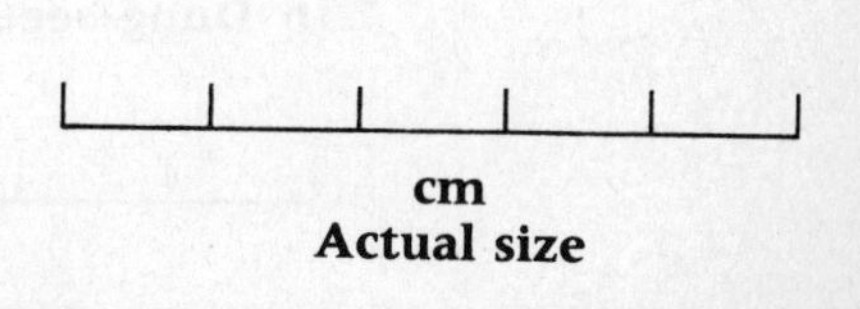

Actual size

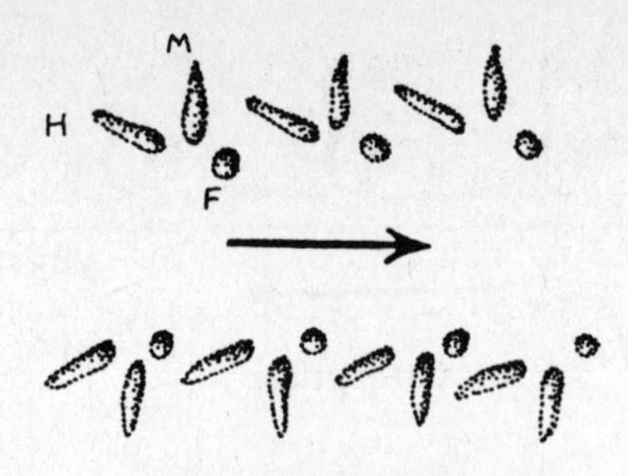

7.3d Rhinoceros beetle

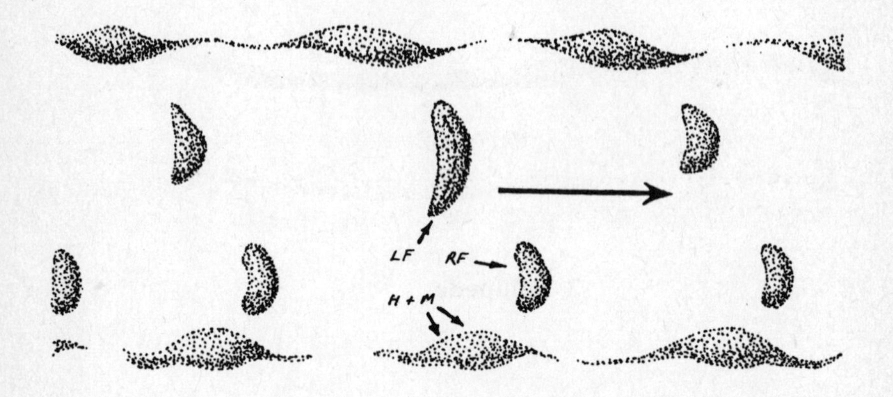

7.3a African dung beetle

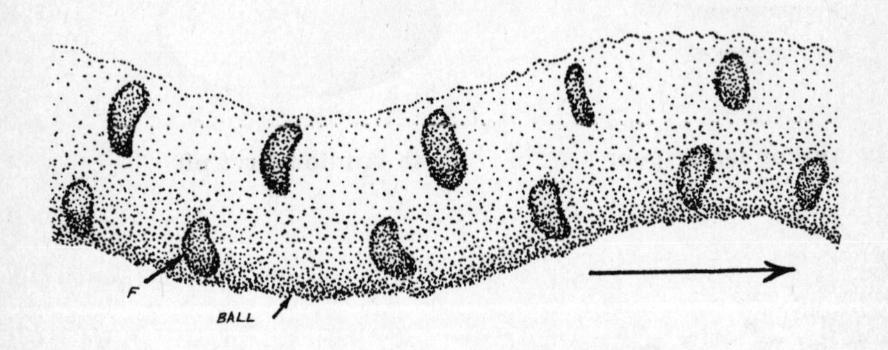

7.3b Dung beetle rolling ball of dung

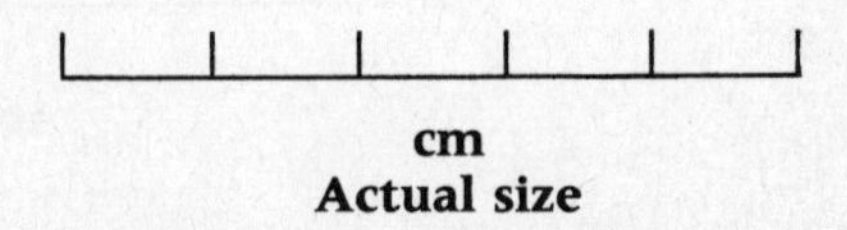

cm
Actual size

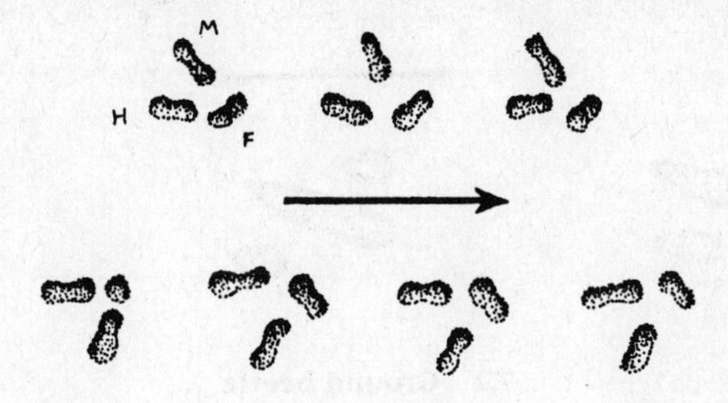

7.1 Tenebrionid beetle (typical spoor)

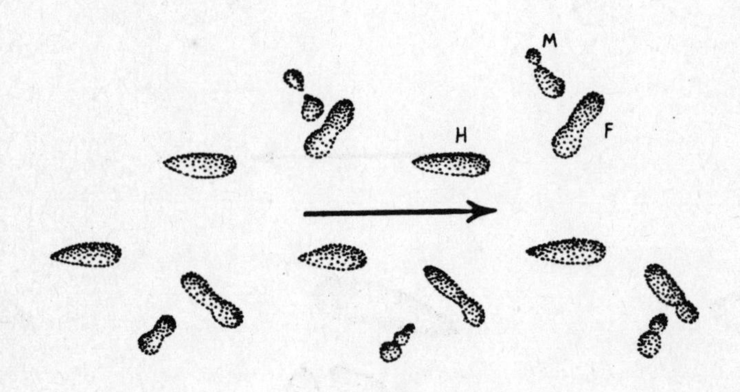

8.1 Grasshopper

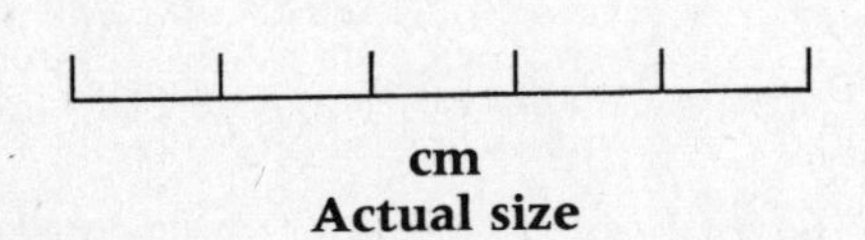

cm
Actual size

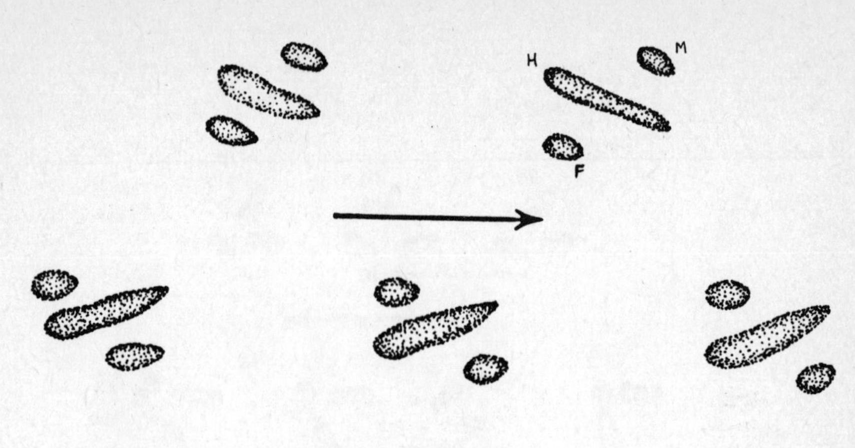

7.2 Ground beetle

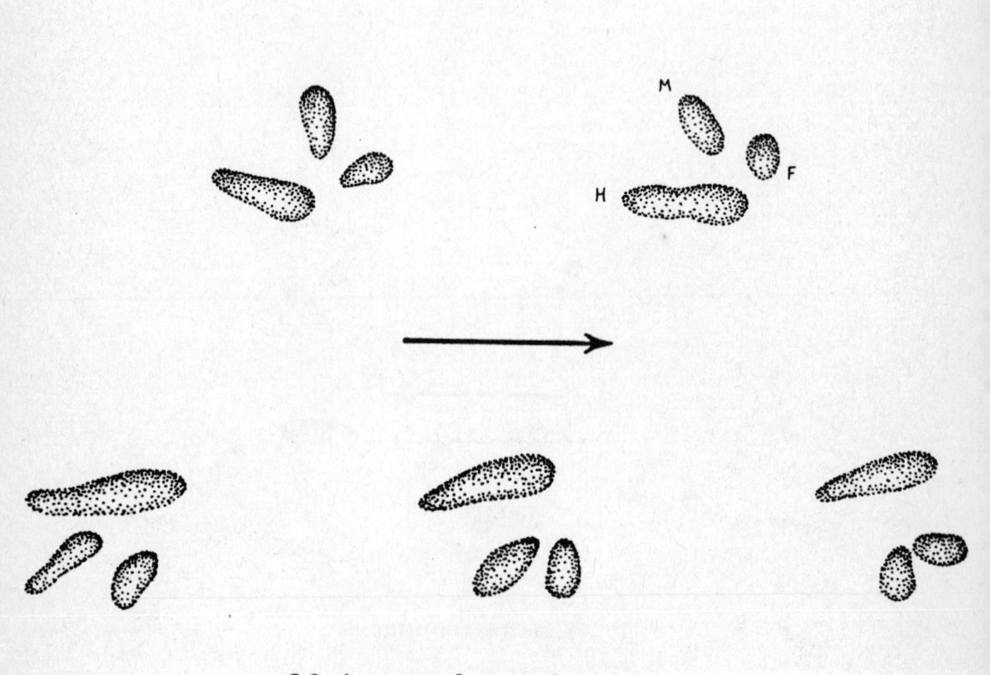

8.2 Armoured ground cricket

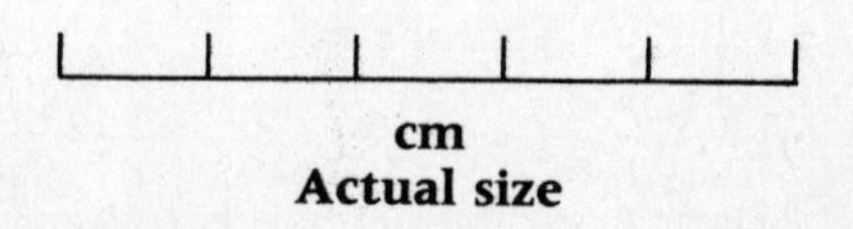

Actual size

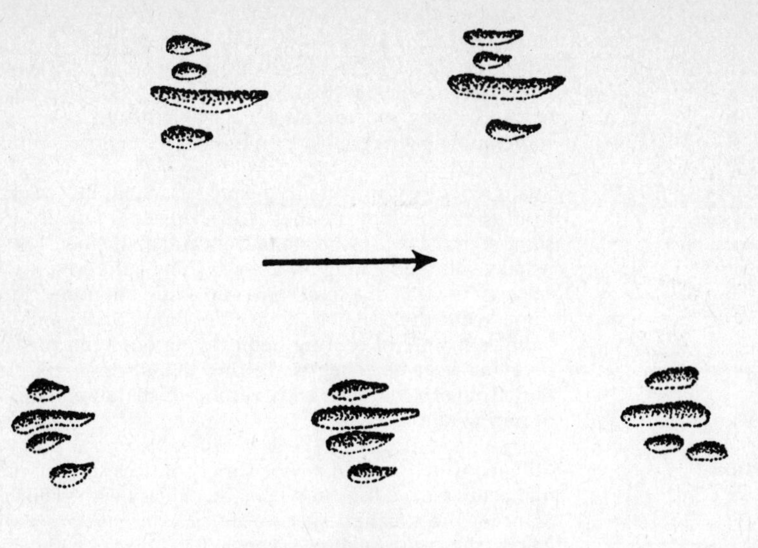

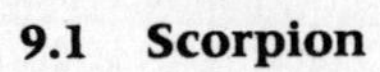

9.1 Scorpion

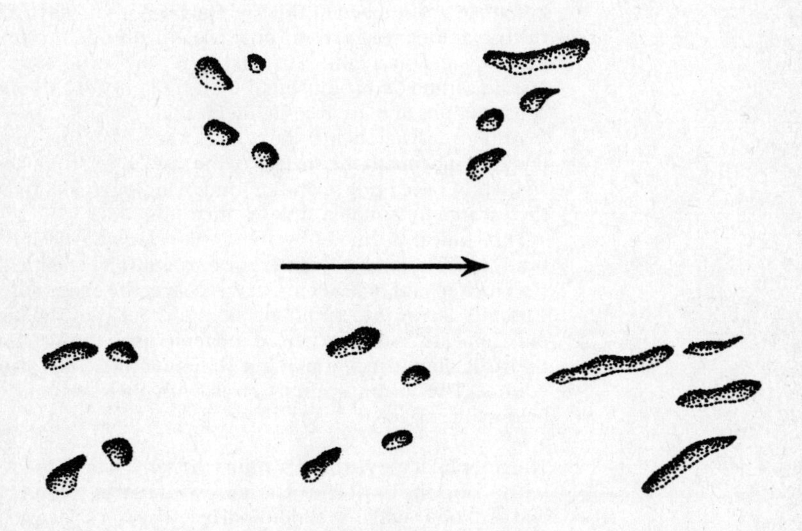

9.2 Hunting spider

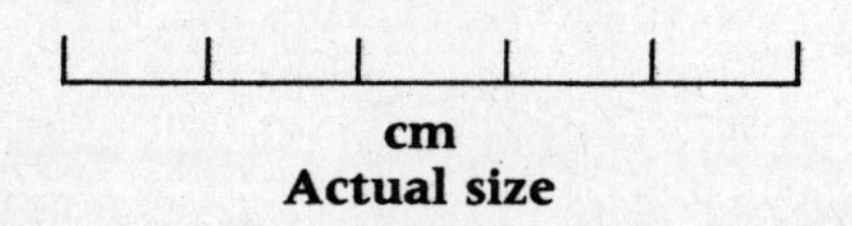

Actual size

Families LUMBRICIDAE & MEGASCOLECIDAE
Earthworms
Erdwurms
1

Earthworms spend most of their time swallowing earth, from which they obtain nourishment in the form of organic material such as decaying plant tissue, seeds, larvae, etc. The castings they produce expose the soil to the air, and the burrows allow air to penetrate the soil, improving drainage and facilitating the growth of roots. They emerge after cold and heavy rain, and may sometimes be seen crawling on the surface in the early morning.

Order STYLOMMATOPHORA
Land snails and slugs
Slakke en naakslakke
2.1 & 2.2

Snails move by waves of muscular contractions moving from the front to the back of the foot. The snail slides forward over the slime secreted by a gland at the front end of the foot. The slime is usually still visible long after it has dried, and presents a shiny surface. The trail of a snail shows up as discontinuous patches of slime, while the trail of a slug is continuous. Snails and slugs are animals mainly of retiring habits living on green plants or on decaying vegetable debris. During the daytime they remain buried out of sight under leaves, from which they emerge at night or during rain.

Class DIPLOPODA
Millipedes
Duisendpote
3

Millipedes are found in a wide variety of sizes and colours. Most millipedes are reddish-brown, black, or black with yellow stripes. Some of the Kalahari species of *Triaenostreptus* are as long as 25 cm; they can usually be seen walking on the open veld after a shower of rain. When disturbed millipedes may writhe vigorously or roll themselves up into tight spirals. They are normally found in moist places. They feed on wood and leaves which have been softened by decay, but certain kinds also feed on living vegetation.

Family MYRMELEONTIDAE
Antlions
4

The adults are like dragonflies in general appearance, but their antennae are clubbed at the tip. They spend the daylight hours resting among vegetation, and take to the air mostly in the evening and at night. The larvae of the various species of *Myrmeleon* and *Cueta* build pits, but the majority of antlions do not construct pits and are free-living in sand.

Antlion larvae always move backwards, tail first, with their bodies just beneath the surface of the sand. Their trails are visible as slightly raised ridges on the sand, winding in all directions as they search for suitable sites for their pits.

The conical pit made by the antlion larvae is found in dry, sandy soil. The larvae lie at the bottom of the pit beneath the sand. If an unwary ant or other insect walks over the edge of the pit and stumbles down the crumbling sides, the head of the antlion at once appears. With jerky head movements, it throws sand up at the struggling insect, making it slip down the side towards the bottom. The victim is then seized with the jaws and dragged below the surface.

Caterpillars
Ruspers
5

The caterpillar moves in very much the same way as earthworms, but anchors the front end of its body with its true legs and the rear end of its body with the abdominal feet. The track it leaves usually shows the prints of the pair of abdominal feet on the last segment. Moths and butterflies begin their active lives as caterpillars. Most caterpillars feed on the juicy parts of leaves, stems, roots or fruit, but some may feed on and live in seeds or woody stems, a few feed on wax and honey, others on wool and hair.

Family FORMICIDAE
Ants
Miere
6

In very fine, soft dust it is sometimes possible to see the footprints of large ants. In coarser sand, where the individual footprints are not visible, ant paths may be seen. All ants are social insects and live in colonies, small or large. A colony consists of one or more queens and a number of workers which are all sterile females. Males are encountered in an ants' nest only at certain seasons of the year when they are being reared for the nuptial flight. As soon as the wedding flight is over the males die, and the queens shed their wings. All ants go through the four stages of egg, larva, pupa and adult. Worker ants are wingless.

Family TENEBRIONIDAE
Tenebrionid beetles (Darkling beetles)
7.1

The typical trail of tenebrionid beetles is characterised by groups of three footprints on either side as shown in Fig. 7.1. This large family, which includes the 'tok-tokkies', contains many thousands of African species whose appearance and habits are diverse and varied. They occur in all types of terrestrial habitats, but the greatest diversity of species is found in dry areas and deserts. The wingless 'tok-tokkies' derive their name from their habit of knocking on the ground loudly at intervals, apparently to attract the opposite sex.

Family CARABIDAE
Ground beetles
7.2

This family contains over 25 000 described species, which are divided in numerous subfamilies. They cannot fly and may be seen running about swiftly during the day. They are fierce hunters with strong, sharp jaws, and can inflict a nasty bite. Their main defence chemical is formic acid, which they can squirt up to 35 cm in any direction when threatened. The fluid can cause severe pain if it comes in contact with human skin, and more serious problems if it gets into the eyes.

Family SCARABAEIDAE
Scarabaeid beetles
7.3

The family Scarabaeidae contains groups such as dung beetles, rose beetles, rhinoceros beetles, chafers and many others. Some of them roll their food into a ball and bury it for themselves and their larvae.

The beetle makes a ball by patting and pressing the dung with its front legs. The spoor of a dung beetle rolling its ball of dung is shown in Fig. 7.3b. It pushes the ball with its middle feet and hind-feet, while walking backwards on its fore-feet. The 'rhinoceros beetles', subfamily Dynastinae, live in decaying wood and plants. Only the males possess the large horns, which are apparently used as weapons in fights between males.

Family ACRIDIDAE
Locusts & grasshoppers
Sprinkane & grassprinkane
8.1

A characteristic feature of grasshopper spoor is that the imprints of the hind-legs are parallel to the direction of motion. The term 'locust' applies to those members of the family Acrididae that gather in swarms and migrate from place to place. The majority of the about 10 000 different species of this family are generally called 'grasshoppers'. Most can jump well and fly strongly. Many have well-developed spines on their powerful hind-legs that are used for self-defence.

Armoured ground cricket
Koringkriek
8.2

The spoor of the armoured ground cricket is fairly large compared with those of other insects. These large insects are widespread, but most abundant in the more arid areas of southern Africa. They are omnivorous, feeding on plant and animal material. Nocturnal in habit, the males start singing after dark with a loud, continuous and piercing buzz. They can inflict a sharp bite with their powerful jaws.

Order SCORPIONES
Scorpions
Skerpioene

9.1

The trail of scorpions is characterised by four tightly grouped footprints on either side. There are two important families in Africa, the Scorpionidae and the Buthidae, which can be distinguished by looking at the pincers and tails. Scorpionidae have large, powerful pedipalps and a slender tail with a small sting. They do not have a powerful venom, and capture and subdue their prey with the pedipalps. The Buthidae, on the other hand, have slender pedipalps and a thick tail with a big sting. They have powerful venom which affects the nervous system, and they kill or paralyse their prey by stinging. A number of Buthidae are dangerous to humans, and their stings can be fatal to some individuals, unless they are treated in time with anti-venom. Scorpions are entirely carnivorous and feed upon any other arthropod. They are active at night and hide during the day.

Order ARANEAE
Spiders
Spinnekoppe

9.2

The trail of a spider shows four footprints on either side. Compared with scorpion spoor, spider footprints are more spread out. Spiders have spinnerets at the posterior and at the abdomen. They use silk for constructing webs, snares, shelters and egg sacs. All spiders are carnivorous, the majority feeding on insects. They hunt at night, and during the daylight hours they seek shelter under stones, the bark of trees, in tangled herbage and in holes in the ground. The only spider in Africa known to be dangerous to humans is the Black Widow Spider, or 'button spider', *Latrodectus mactans.* It has a round, shiny black abdomen with a crimson line or spot on the dorsal surface.

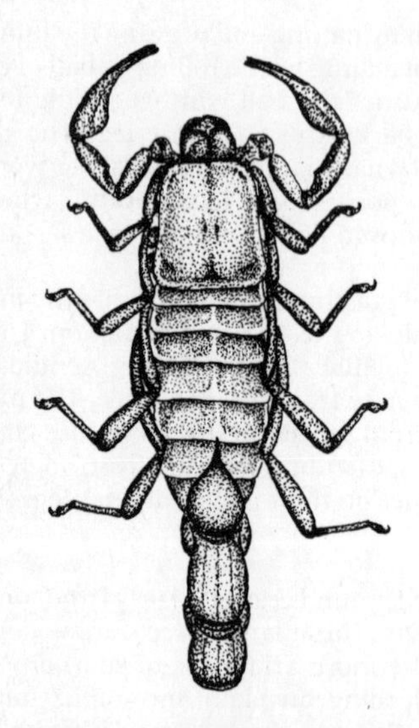

Scorpion, family Buthidae

WATCH IT
THIS IS THE BADDIE!
SKINNY

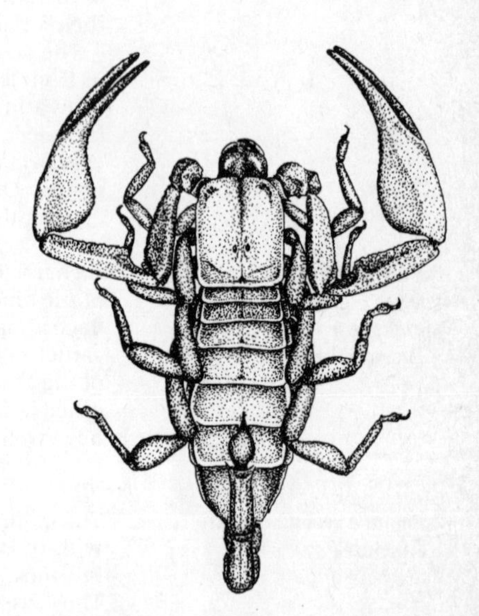

Scorpion, family Scorpionidae

AMPHIBIANS
&
REPTILES

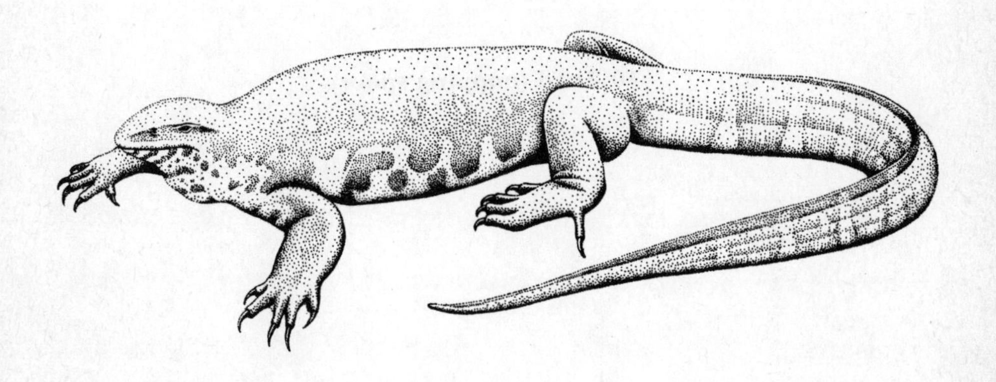

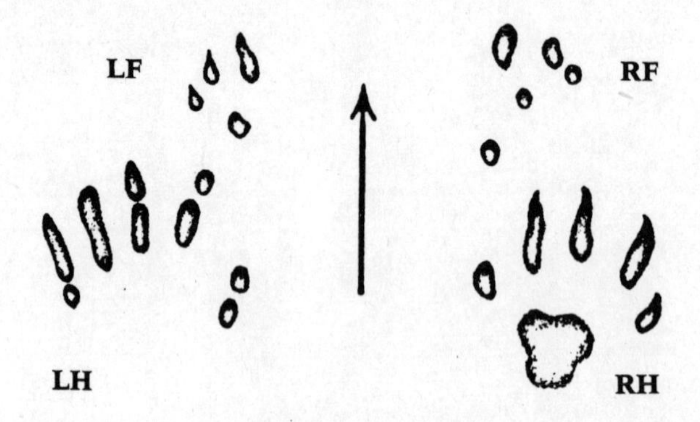

10 Toad (hopping spoor)

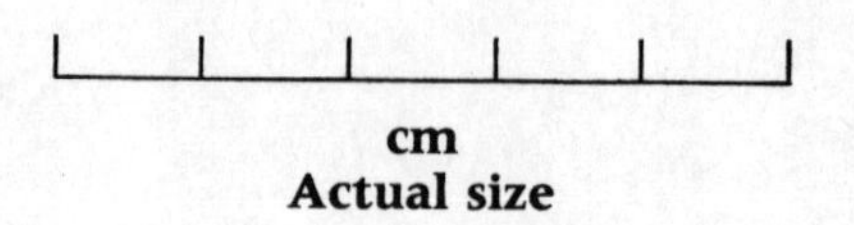

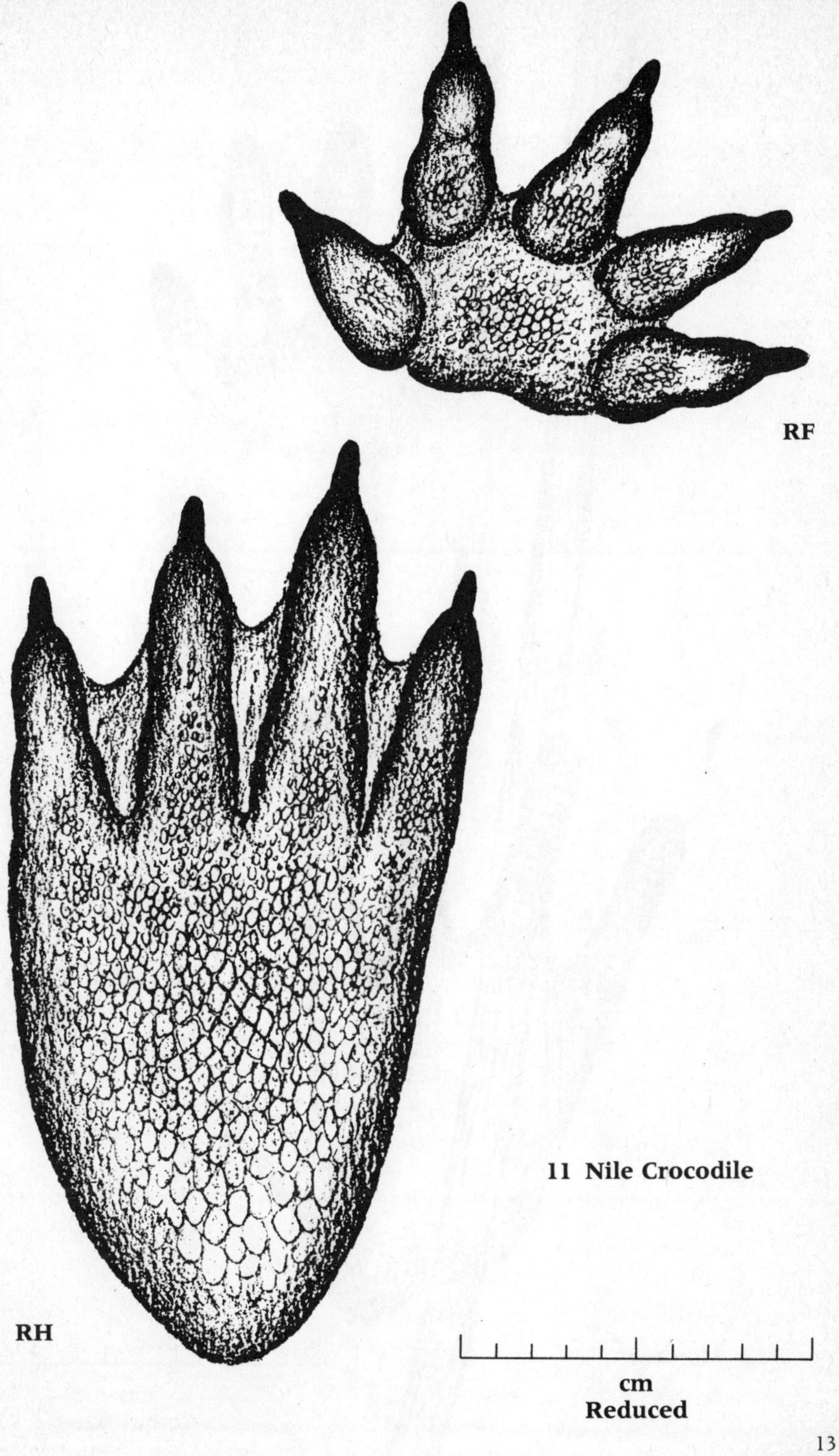

11 Nile Crocodile

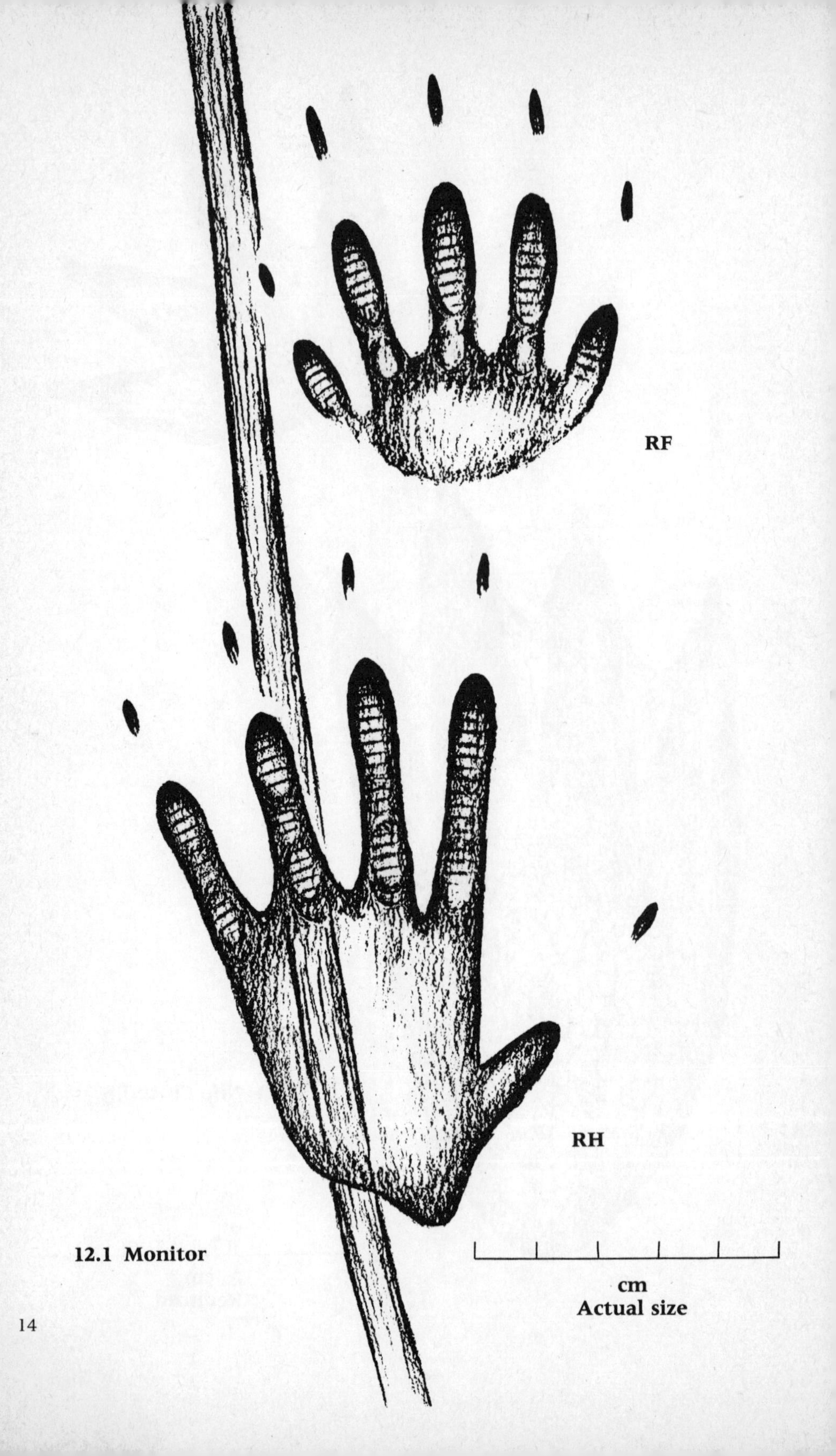

12.1 Monitor

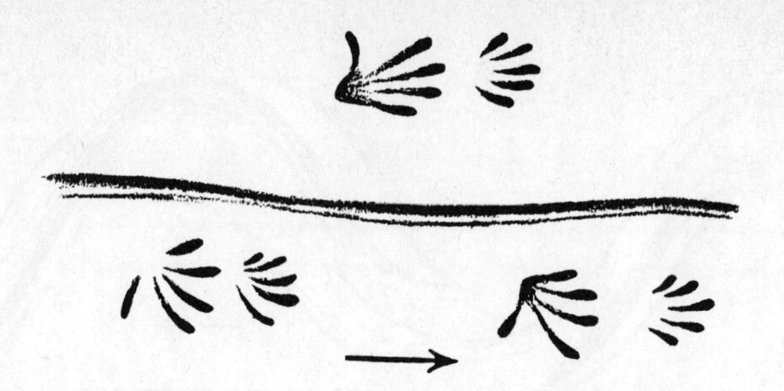

12.2a Skink

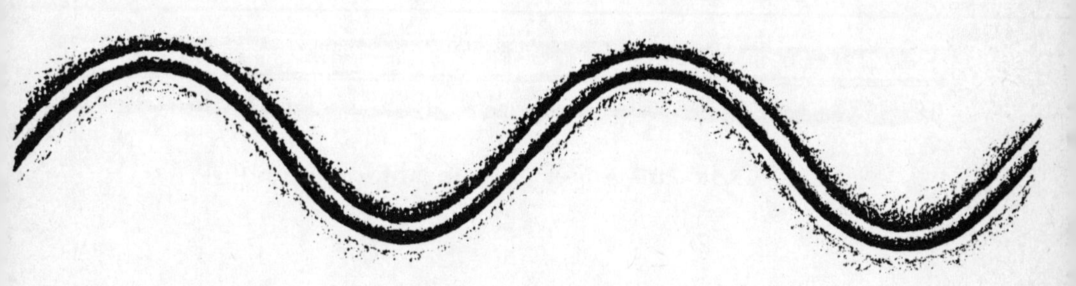

12.2b Legless Skink

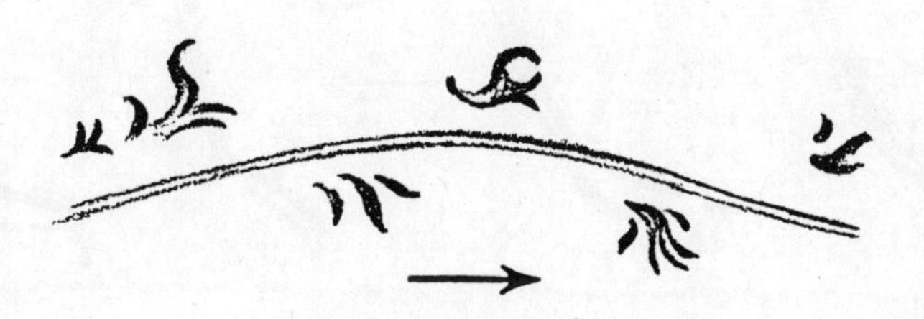

12.3 Lizard

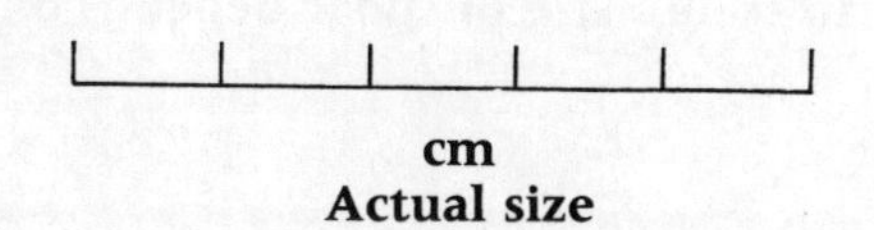

cm
Actual size

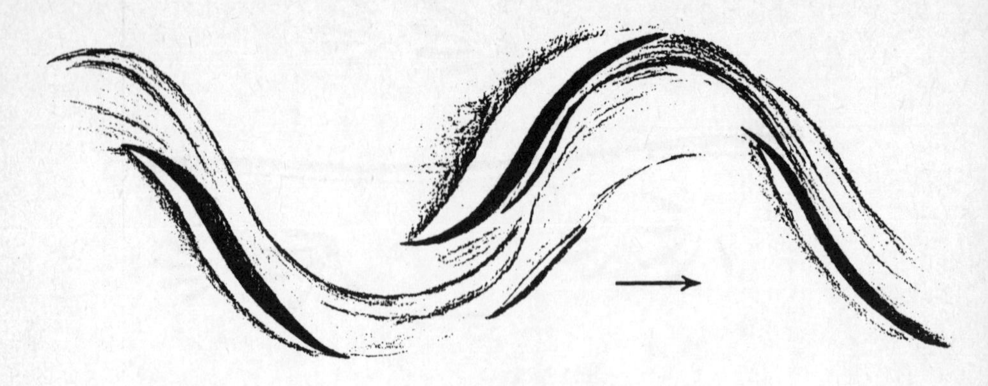

13.1 Typical snake

13.3a Puff Adder (rectilinear locomotion)

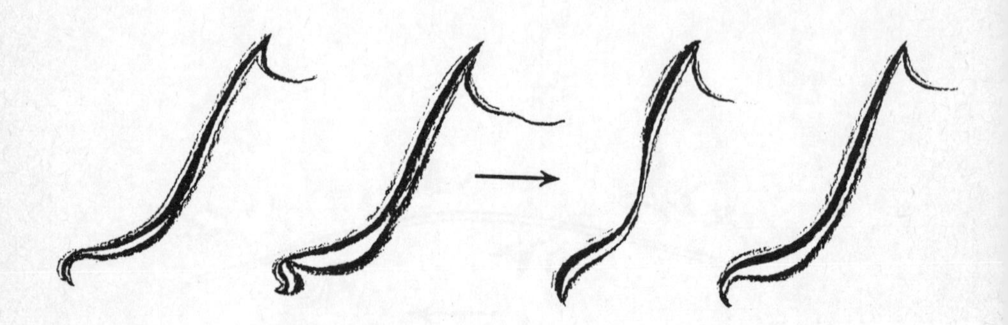

13.3c Single-horned Adder (sidewinding)

Reduced (not to scale): size of spoor depends on length of snake

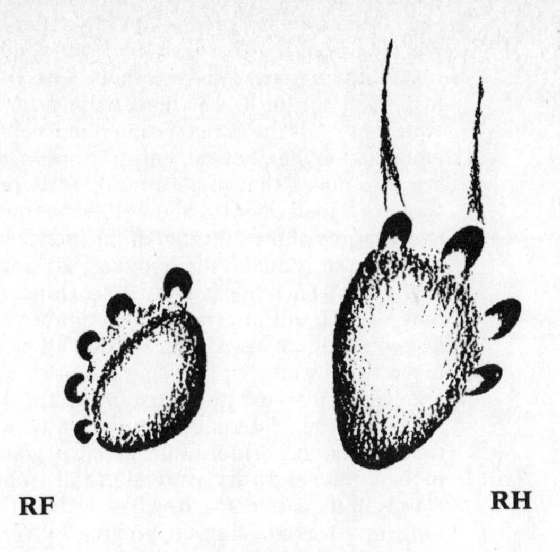

14.1a Angulate Tortoise

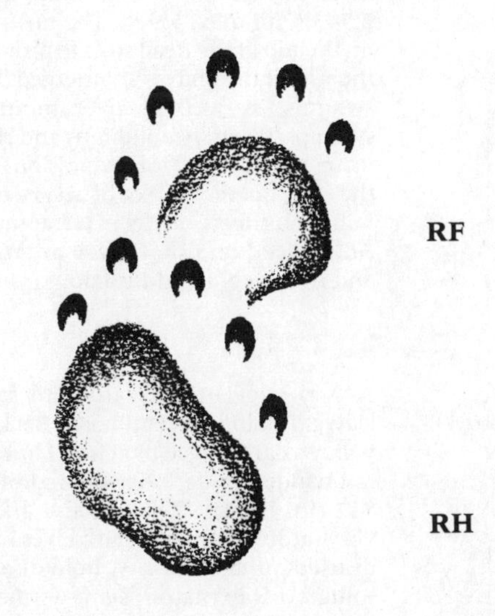

14.5 Side-necked Terrapin

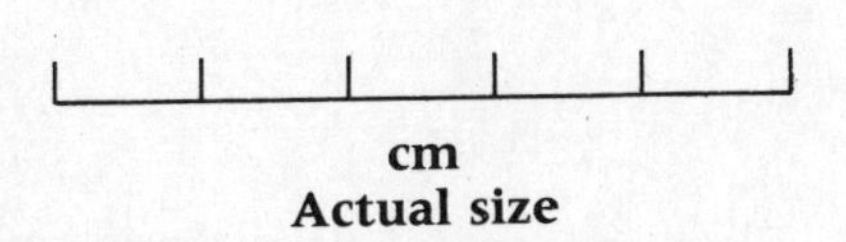

Order ANURA
Frogs & toads
Paddas & brulpaddas

10

Frogs and toads have four toes on each fore-foot and five toes on each hind-foot. In predominantly aquatic forms, which have webbing between the toes, the lever-system type of hind-limb makes them powerful swimmers. The frog is adapted to the leaping mode of locomotion, mainly for leaping to safety in water. In toads the skeletal components of the hind-limb are less elongated so that they can only hop or run, while in some the legs are so reduced that they can only walk or crawl. The hopping spoor of a toad (Fig. 10) shows the five toes of the hind-feet and the imprints of the four toes of the fore-feet. The irregular shape of the spoor is due to the hopping, which causes the toes to slip in the sand. Each hop is about 30 cm long, although the distance may vary. Toads in general inhabit more open country and are less aquatic than most frogs, although they are dependent on the presence of water for breeding purposes. Like most frogs, toads are crepuscular and nocturnal in habit.

The obvious differences between a typical frog and a typical toad are a smooth, moist skin devoid of glandular concentrations in the former and a dry, warty skin and usually prominent parotid glands in the latter. The frog has teeth in the upper jaw and has long hind-legs enabling it to progress by a series of leaps, whereas the toad is edentulous and has short hind-legs so that it can only walk or hop. The frog is more aquatic, while the toad is more terrestrial. Both frogs and toads eat insects, insect larvae and snails.

Crocodylus niloticus
Nile Crocodile
Nylkrokodil

11

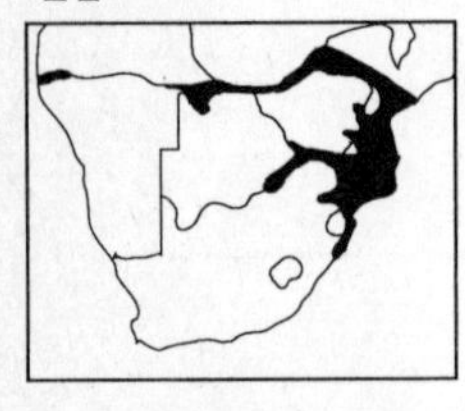

Nile Crocodiles may exceptionally exceed 1 000 kg. TL 2,5–3,5 m; max. 5,9 m. The nostrils, eyes and ear-openings are on the top of the head so as to project slightly when floating while the rest of the body is submerged. **Habitat:** Larger rivers, lakes and swamps, as well as river mouths, estuaries and mangrove swamps. **Habits:** Amphibious and riparian, it is a strong swimmer, using only the tail for propulsion. Much time is spent basking in the sun on the banks of rivers and the shores of lakes. **Food:** Subadults feed on fish, terrapins, birds and small mammals. Adults feed on fish, as well as large mammals such as antelope and even zebra and Buffalo.

Varanus exanthematicus
Rock, Tree or White-throated Monitor
Veldlikkewaan

12.1a

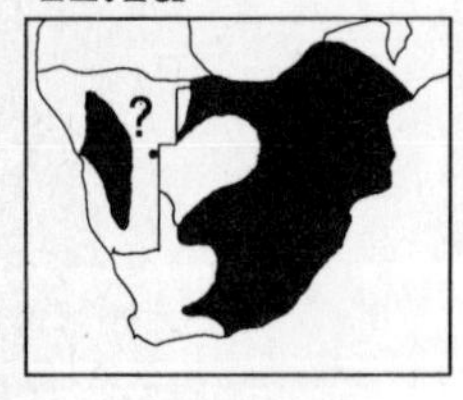

A very large, stout lizard with strong, stocky limbs and sharp claws. Tail longer than body. Back is dark grey-brown with pale yellow, dark-edged blotches. Limbs spotted with pale yellow and tail banded in dark brown and off-white. TL 70–110 cm; max. 132 cm. **Habitat:** Savanna and arid, karroid areas. Often found very far from water. **Habits:** Lives in tunnel under rock overhang, disused animal burrow, hole in a tree or a rock crack. Usually solitary. Hibernates, semi-dormant in winter. **Food:** Mainly invertebrates although it will kill and eat any animal small enough to swallow. Also scavenges.

Varanus niloticus
Water or Nile Monitor
Waterlikkewaan

12.1b

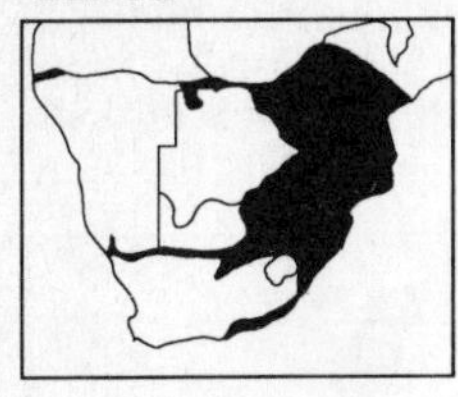

The largest African lizard, it has a stout body, powerful limbs and strong claws. Tail much longer than body. Greyish-brown to dirty olive-brown on top of head and back, with scattered darker blotches and light-yellow ocelli and bands on head, back and limbs. Belly and throat paler, with black bars. **Habitat:** Rivers, pans and major lakes. Always found in the vicinity of permanent water. **Habits:** Excellent swimmer. Often basks on rock outcrops or tree stumps. In temperate regions, may hibernate communally. **Food:** Diet varied. Forages in freshwater pools for crabs and mussels, as well as frogs, fish, birds and eggs.

Family SCINCIDAE
Skinks

12.2

When it moves slowly on firm ground with a thin layer of dust, the footprints of a typical skink may show the five toes of the fore- and hind-feet, with the hind footprint just behind the fore footprint (Fig. 12.2a). The tail leaves a slightly wavy line. The spoor of a legless skink, such as the Striped Blind Legless Skink (or Kalahari Blind Skink), *Typhlosaurus lineatus*, is characterised by the neat sinuous pattern created as it 'swims' through loose sand with an undulating progression (Fig. 12.2b). Although most skinks have well-developed limbs, some have only vestiges of limbs or no limbs. The tail can be shed and regenerated. Skinks are for the most part terrestrial, although many climb trees and rocks. They are active during the day, and most maintain a high body temperature by shuttling between sunny spots and shade. They feed almost exclusively on small insects, which they seize after a short rush from cover. Legless skinks are burrowing reptiles and feed on earthworms, beetle larvae and termites.

Family LACERTIDAE
Old World Lizards or Lacertids

12.3

The spoor of a lacertid is similar to that of a skink. In soft sand the footprints have irregular shapes as the long toes of the back feet are swept through the sand (Fig. 12.3). The long tail leaves a thin furrow. These small to medium-sized lizards have slender bodies, long tails and well-developed legs. The dorsal scales are usually small, smooth and granular. The tail has whorls of keeled scales, which may be spiny, and it can be shed and regenerated. They are active, diurnal and mainly terrestrial, although some are rock-living or arboreal.

Family COLUBRIDAE
Typical snakes

13.1

The spoor in soft sand of a medium-sized, slenderly built Typical snake is shown in Fig. 13.1a. When moving at normal speed by means of serpentine progression, it leaves a sinuous trail. This large family of snakes contains some of the most common snakes. Most are medium-sized and harmless, although a few are dangerous, such as the Boomslang, Bird Snake and burrowing asps.

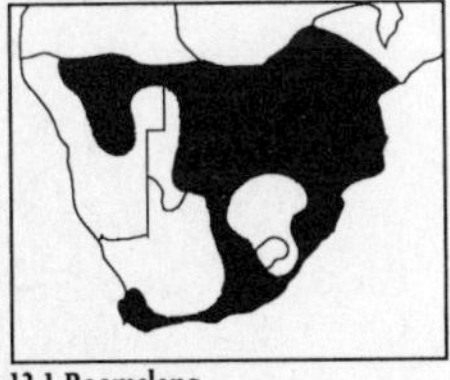

13.1 Boomslang

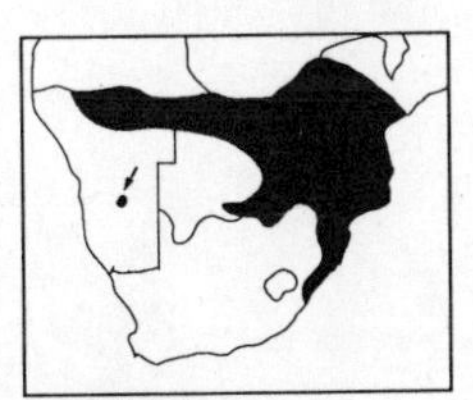

13.1 Bird Snake

Genus *Naja*
Cobras

13.2a

Compared with that of a slenderly built Typical snake, the spoor of the cobra indicates a thicker and more massive body and is similar to the spoor of more stockily built Typical snakes such as the Mole Snake (Fig. 13.1b). The cobras are large, stockily built, terrestrial snakes with smooth scales. They are active hunters, pursuing and capturing small vertebrates. All are potentially dangerous, and when threatened, they lift the forebody and spread their hood.

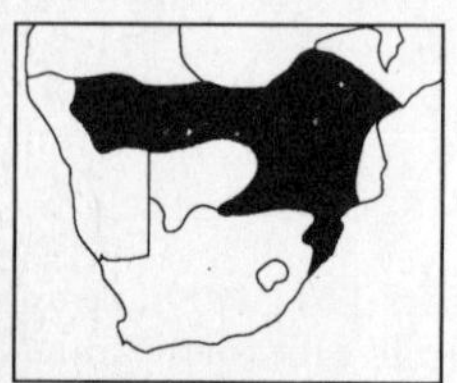

13.2 Egyptian Cobra

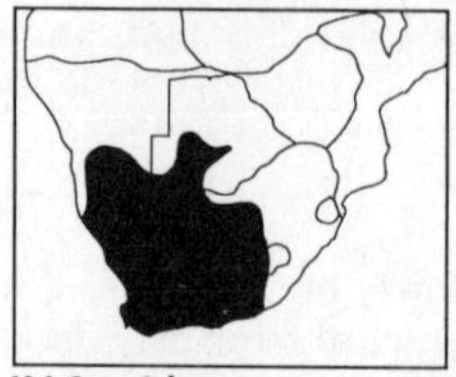

13.2 Cape Cobra

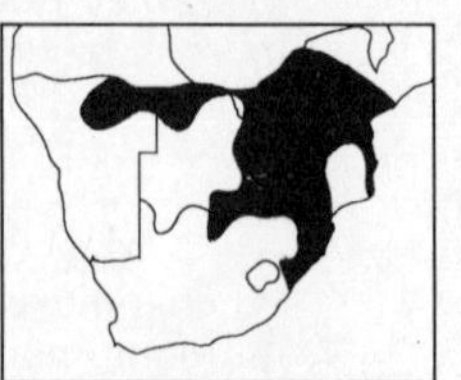

13.2 Mozambique Spitting Cobra

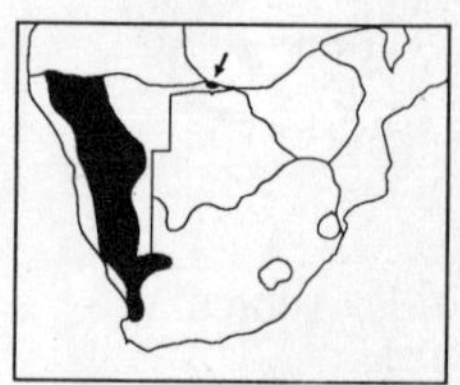

13.2 Black-necked Spitting Cobra

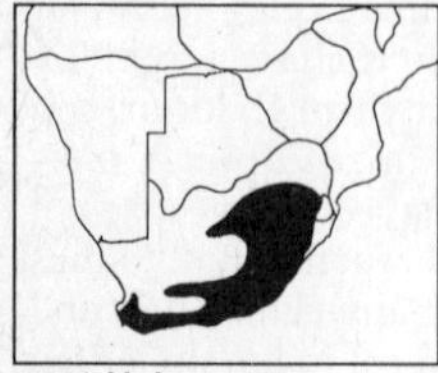

13.2 Rinkhals

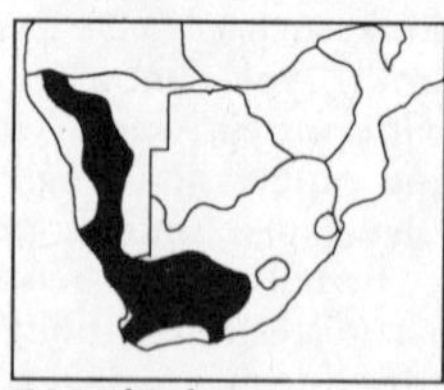

13.2 Coral Snake

Genus *Dendroaspis*
Mambas

13.2b

A characteristic feature of the mamba spoor is the indication of the wide sweeping motions of its long, slender body as it moves by serpentine progression. Two species of mamba are found in southern Africa, namely the Black Mamba and the Green Mamba, *Dendroaspis angusticeps*. These large, agile snakes have long, flat-sided heads. They are diurnal and actively pursue their prey, striking rapidly until it succumbs to the poison. They are dangerous to humans, although only the Black Mamba commonly bites.

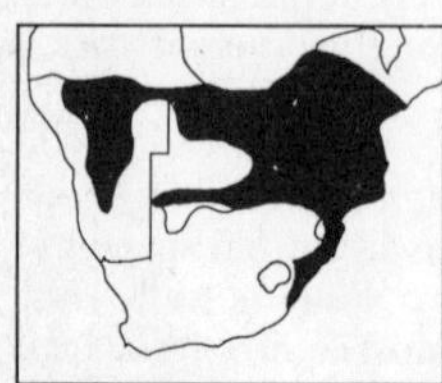

13.2 Black Mamba

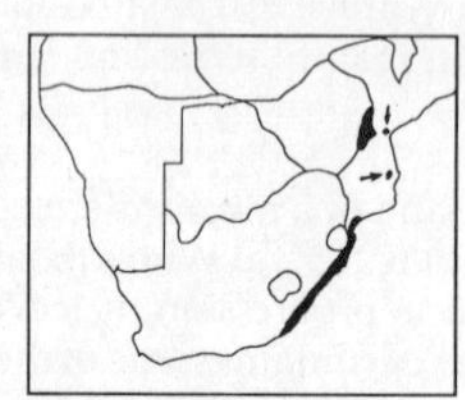

13.2 Green Mamba

Family VIPERIDAE
Adders and vipers

13.3

The Puff Adder, *Bitis arietans,* normally moves by means of rectilinear progression, leaving a straight trail in the sand (Fig. 13.3a). When travelling faster, the Puff Adder moves by means of serpentine progression, leaving an undulatory spoor. Peringuey's Adder, *Bitis peringueyi,* the Horned Adder, *Bitis caudalis,* the Many-horned Adder, *Bitis cornuta,* and the Namaqua Dwarf Adder, *Bitis schneideri,* adopt the sidewinding progression to move swiftly over hot, loose sand. The spoor shows a series of parallel lines, leaving gaps as the snake lifts its body in an undulating motion (Fig. 13.3c). The African adders, genus *Bitis,* are stocky snakes with distinct heads and short tails. They are terrestrial and usually nocturnal or crepuscular. They have large, erectile fangs at the front of the mouth and feed on small vertebrates, which are ambushed and killed by the venom.

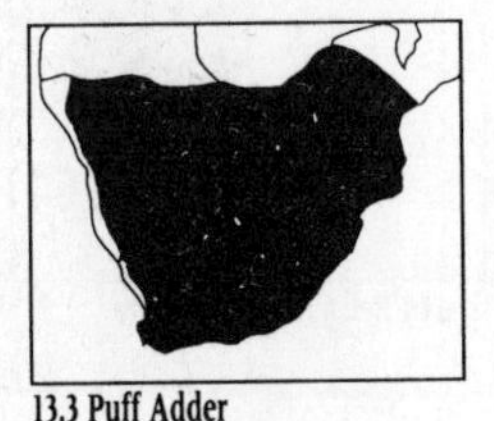

13.3 Puff Adder

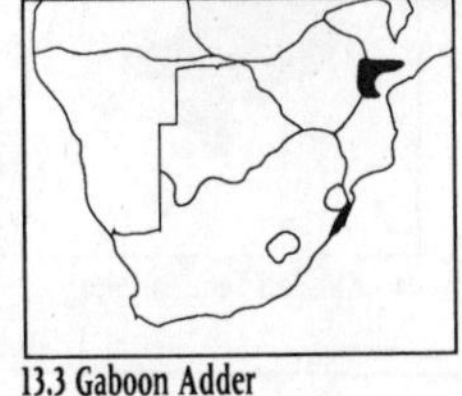

13.3 Gaboon Adder

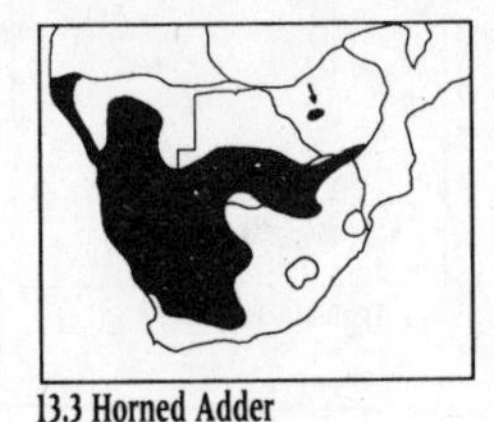

13.3 Horned Adder

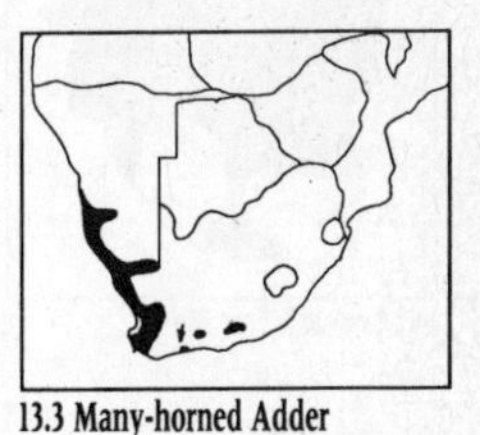

13.3 Many-horned Adder

13.3 Peringuey's Adder

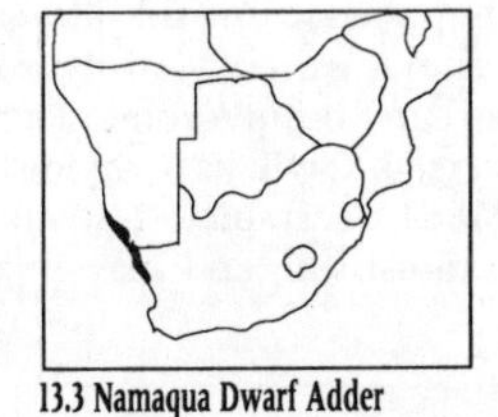

13.3 Namaqua Dwarf Adder

Genus *Python*
Pythons

13.4

When moving slowly, the African Rock Python, *Python sebae,* adopts a rectilinear progression, leaving a straight trail in the sand. When travelling faster the python moves in an undulatory progression, leaving a spoor that shows wide sweeping motions. Pythons are medium to large snakes with small, smooth scales. Two species occur in southern Africa — Anchieta's Dwarf Python, *Python anchietae,* and the African Rock Python, Africa's largest snake. Prey is ambushed and constricted, usually at dusk or after dark. The African Rock Python can swallow very large prey, but is vulnerable when swollen with food. It may attack humans, but such attacks are very rare.

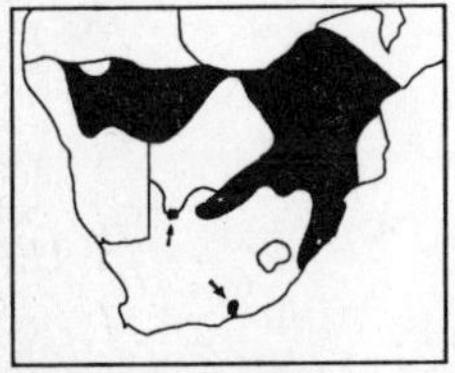

13.4 African Rock Python

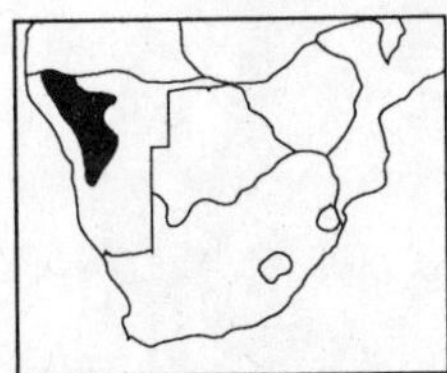

13.4 Anchieta's Dwarf Python

Family TESTUDINIDAE
Land tortoises
14.1

Of the land tortoises, two species of padlopers, the Greater Padloper, *Homopus femoralis*, and the Parrot-beaked Tortoise, *Homopus areolatus*, have four claws on the fore- and hind-feet. All other land tortoises, including three species of padlopers, have five claws on the fore-feet and four claws on the hind-feet. Land tortoises are cold-blooded and cannot tolerate extreme temperatures. In cold periods they shelter underground or in some other sheltered place, while in summer they will seek shade from the midday sun. For protection they withdraw into their hard, bony shell. Tortoises feed mainly on plants, although some may eat invertebrates such as snails and millipedes, or may gnaw bones and even eat hyaena droppings for calcium.

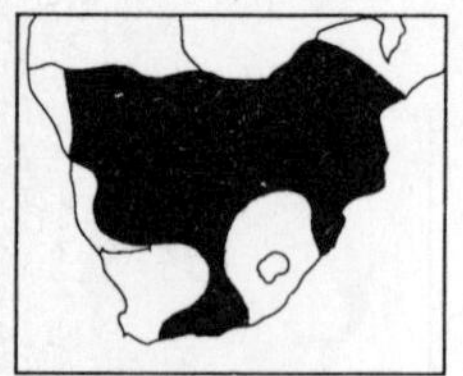
14.1 Leopard Tortoise

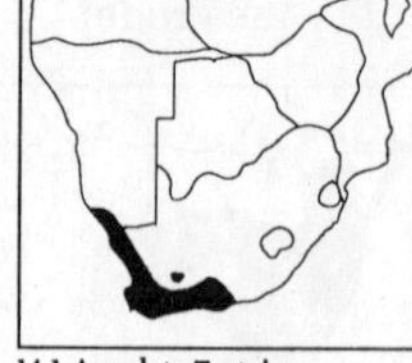
14.1 Angulate Tortoise

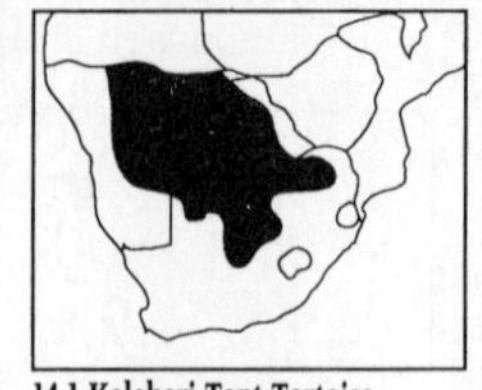
14.1 Kalahari Tent Tortoise

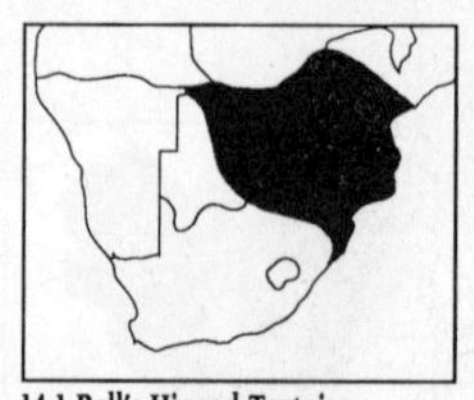
14.1 Bell's Hinged Tortoise

Family PELOMEDUSIDAE
Side-necked terrapins
Subfamily PELOMEDUSINAE
14.5

Side-necked terrapins are characterised by the way they withdraw their heads, pulled to one side under the carapace. The shell is flat and hard. They are mainly carnivorous, while some, like the Marsh Terrapin, are omnivorous. They live in pans, lakes, swamps and rivers, and will bask on logs, rocks or banks. Some, such as the Marsh Terrapin, aestivate during droughts by burrowing into moist soil, and may migrate to new vleis after good rains.

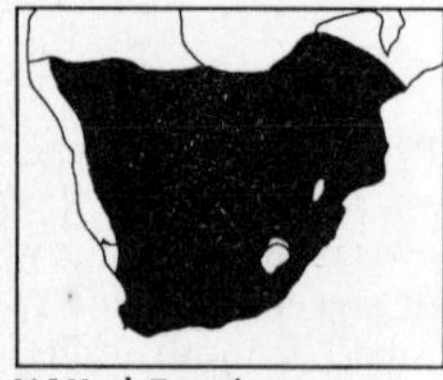

14.5 Marsh Terrapin

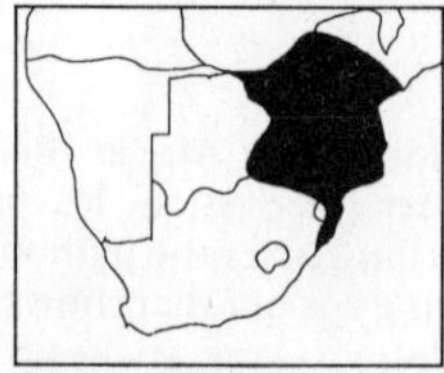
14.5 Serrated Hinged Terrapin

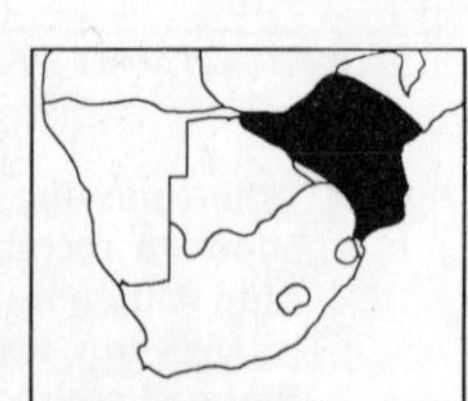
14.5 Pan Hinged Terrapin

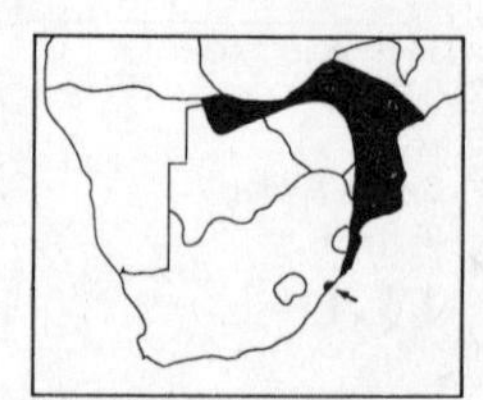
14.5 Mashona Hinged Terrapin

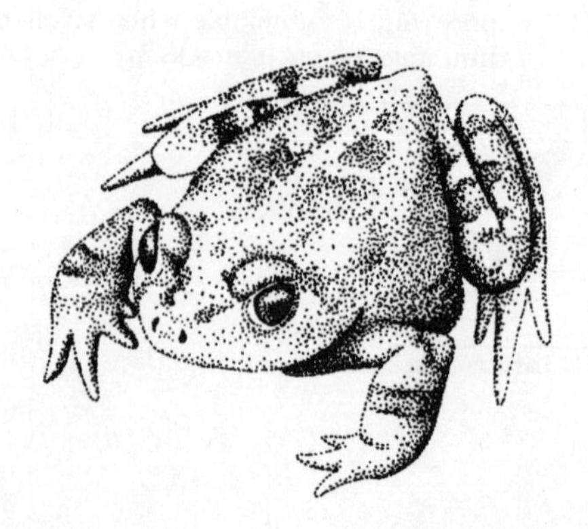

BIRDS

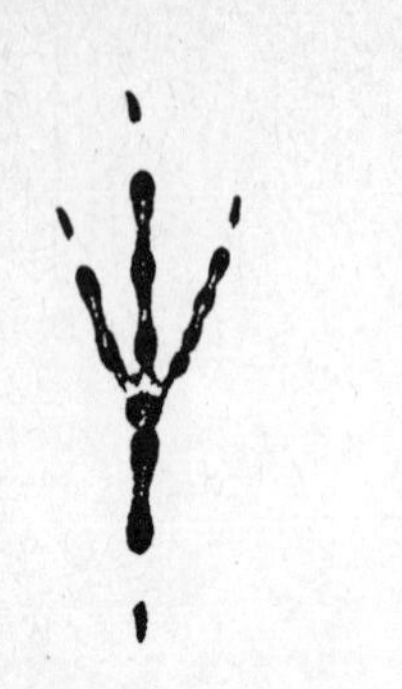

15.1 Small passerine, family Ploceidae

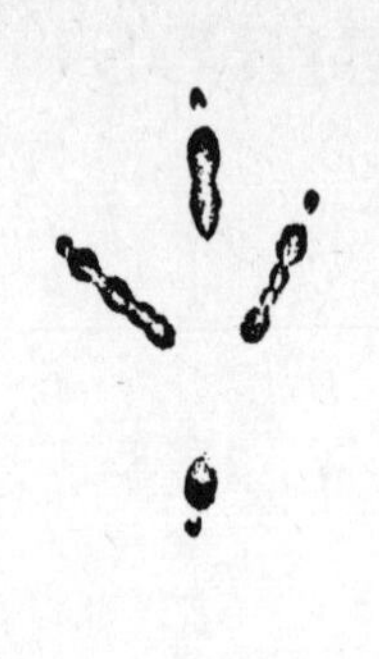

26.1 Common Quail

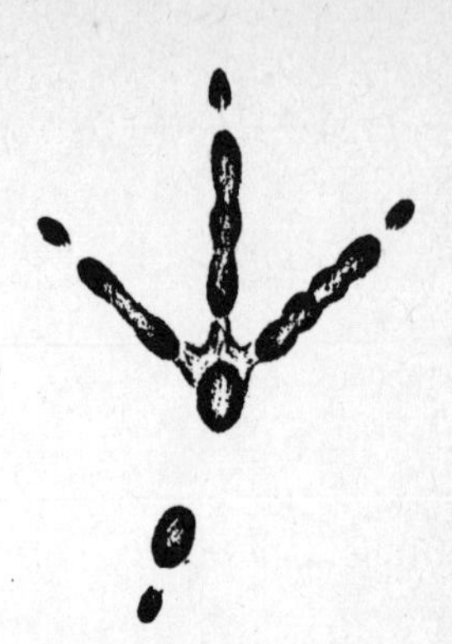

19.3 Cape Turtle Dove

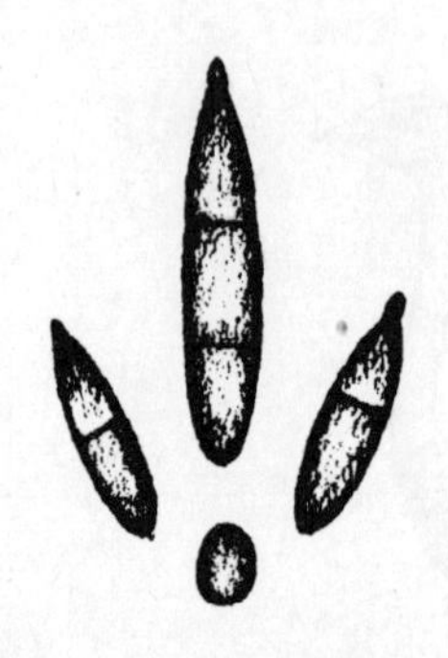

23.1 Spotted Dikkop

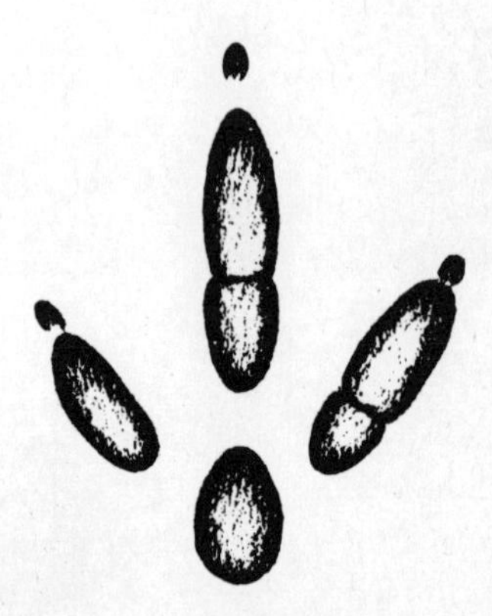

30.1 Black Korhaan

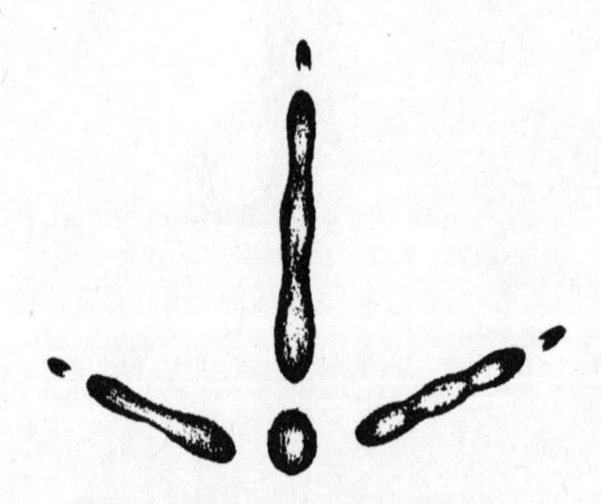

22.2 Blacksmith Plover

cm
Actual size

Caterpillar and its tracks

Dung beetles

Marsh Terrapin

Scorpion, family Buthidae

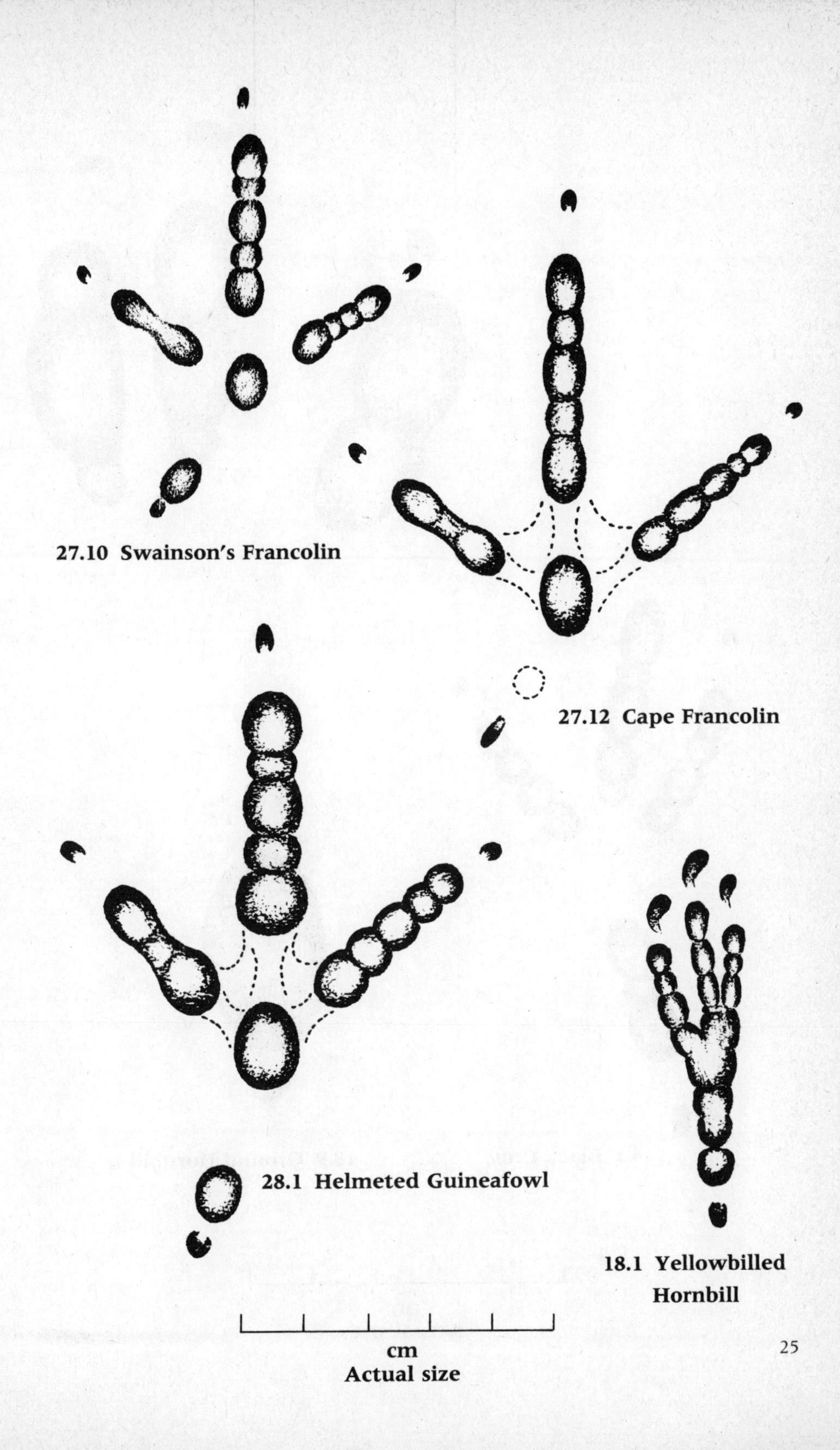

27.10 Swainson's Francolin

27.12 Cape Francolin

28.1 Helmeted Guineafowl

18.1 Yellowbilled Hornbill

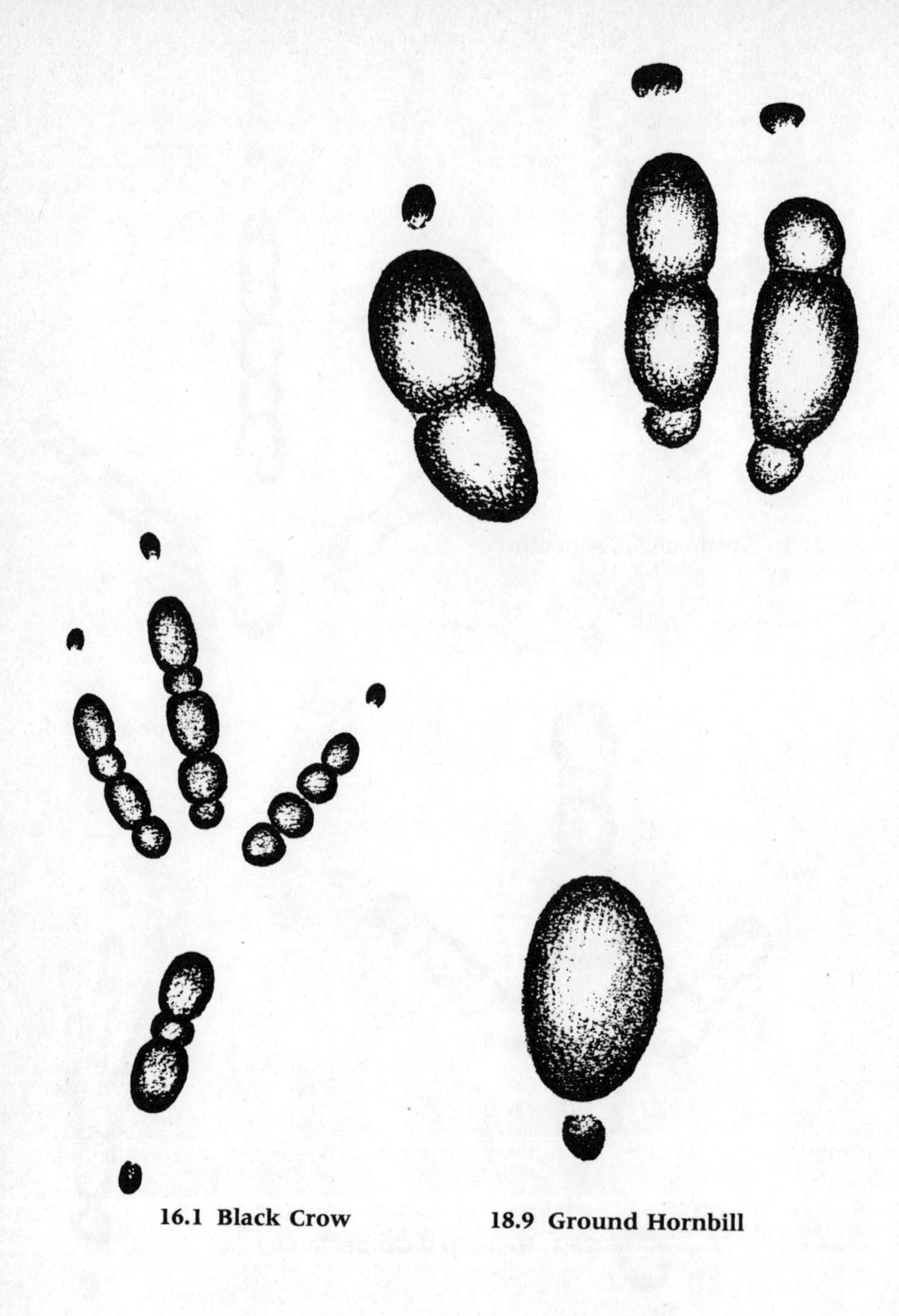

16.1 Black Crow **18.9 Ground Hornbill**

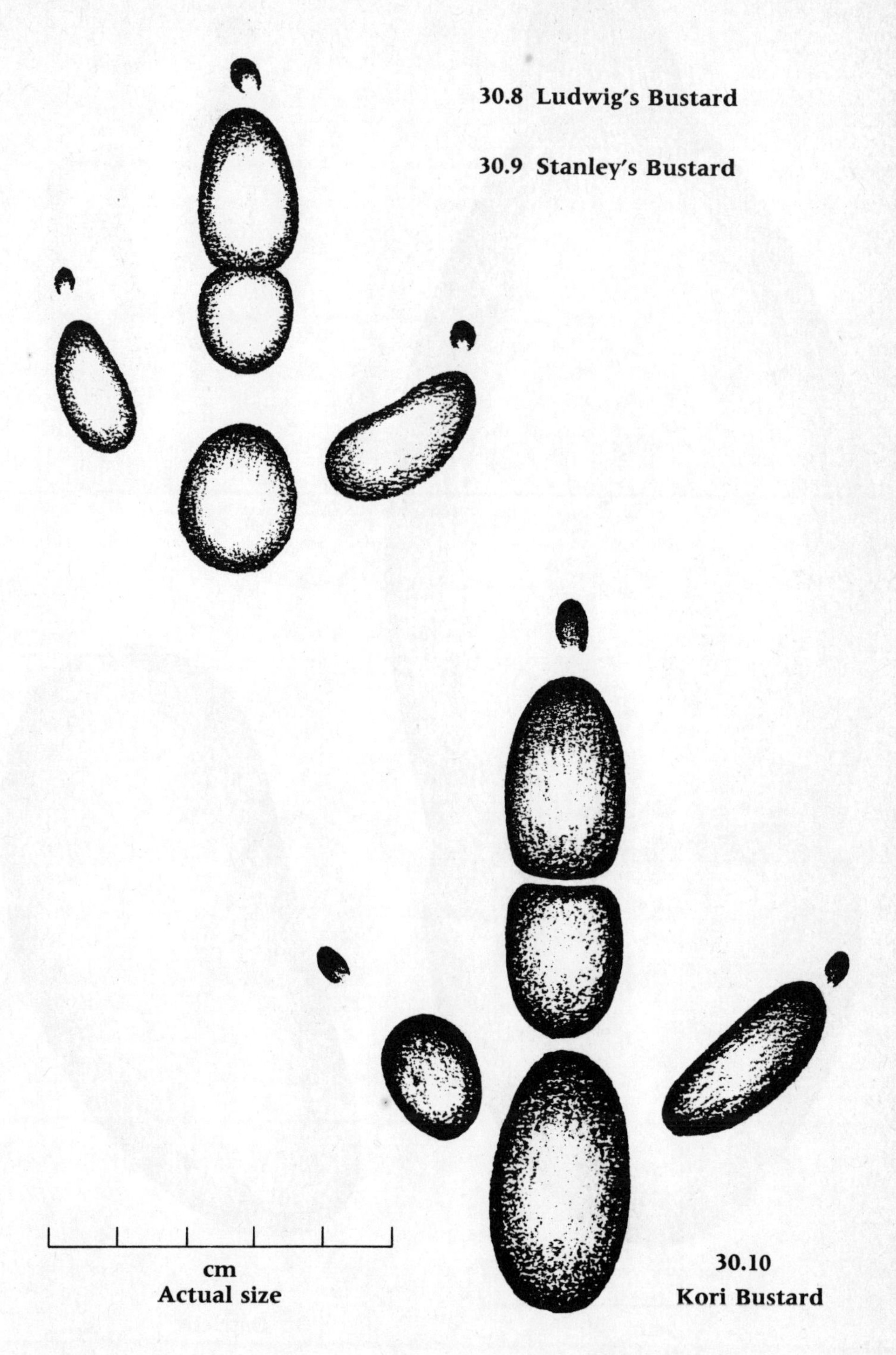
30.8 Ludwig's Bustard
30.9 Stanley's Bustard
cm
Actual size
30.10
Kori Bustard

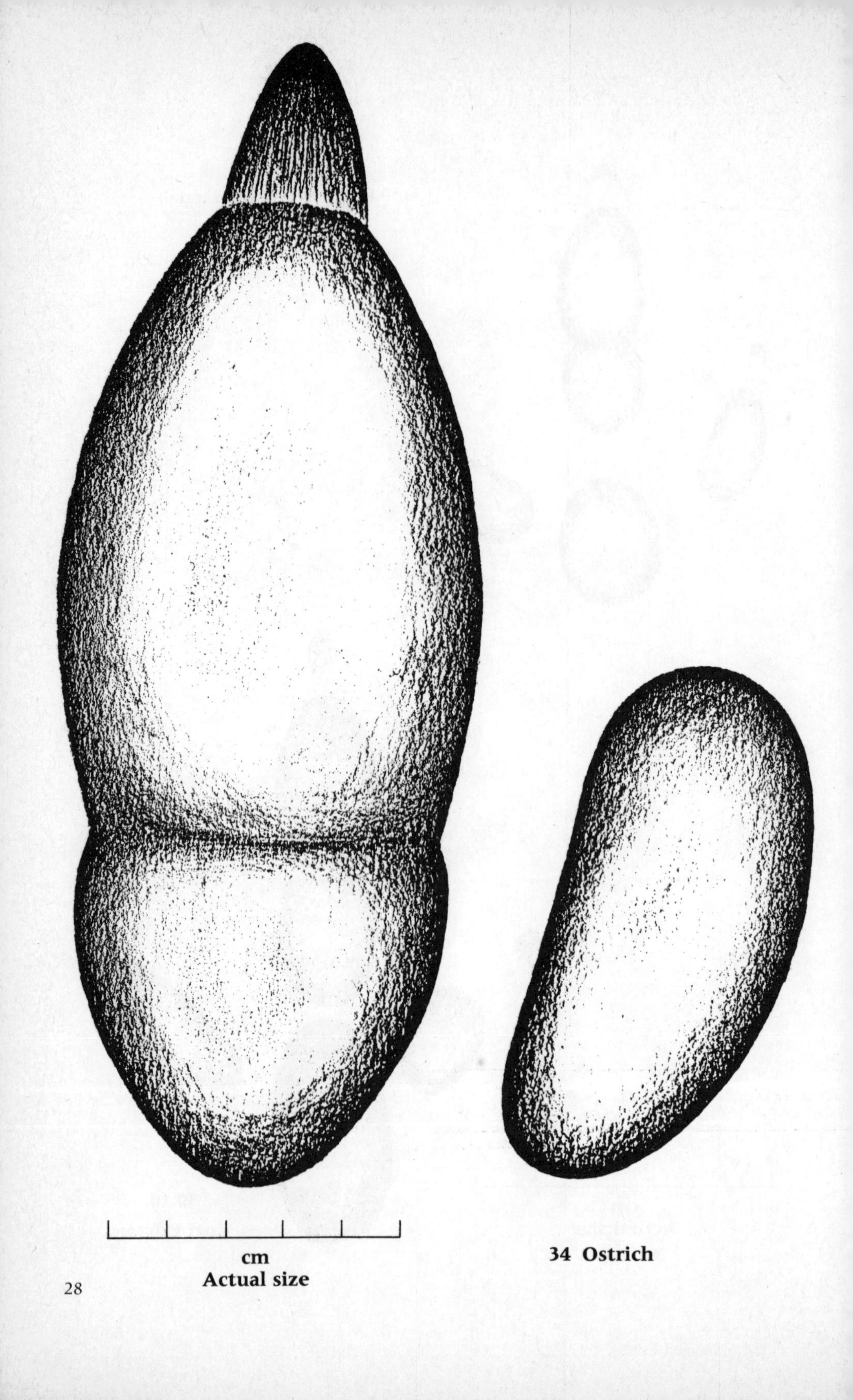

34 Ostrich

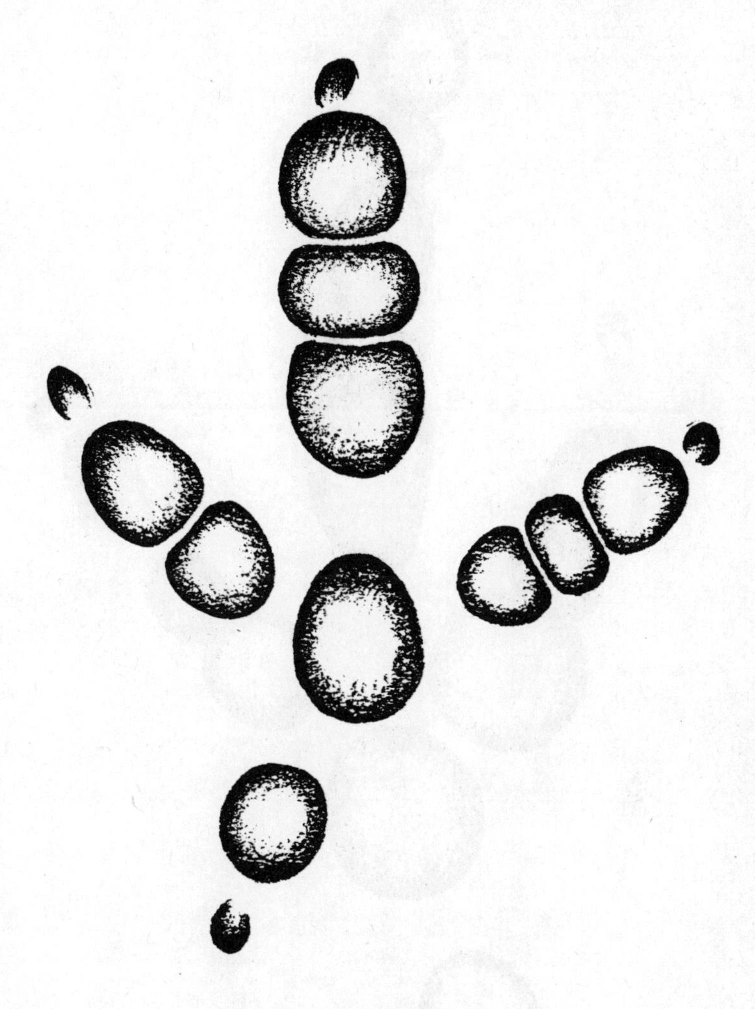

35 Secretarybird

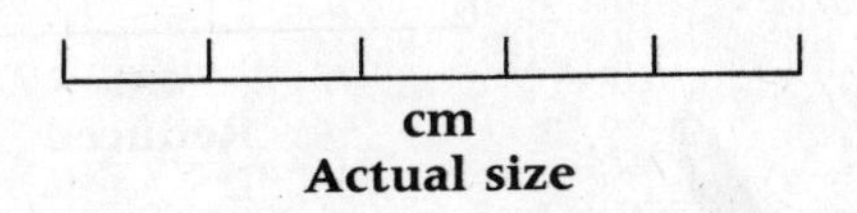

Actual size

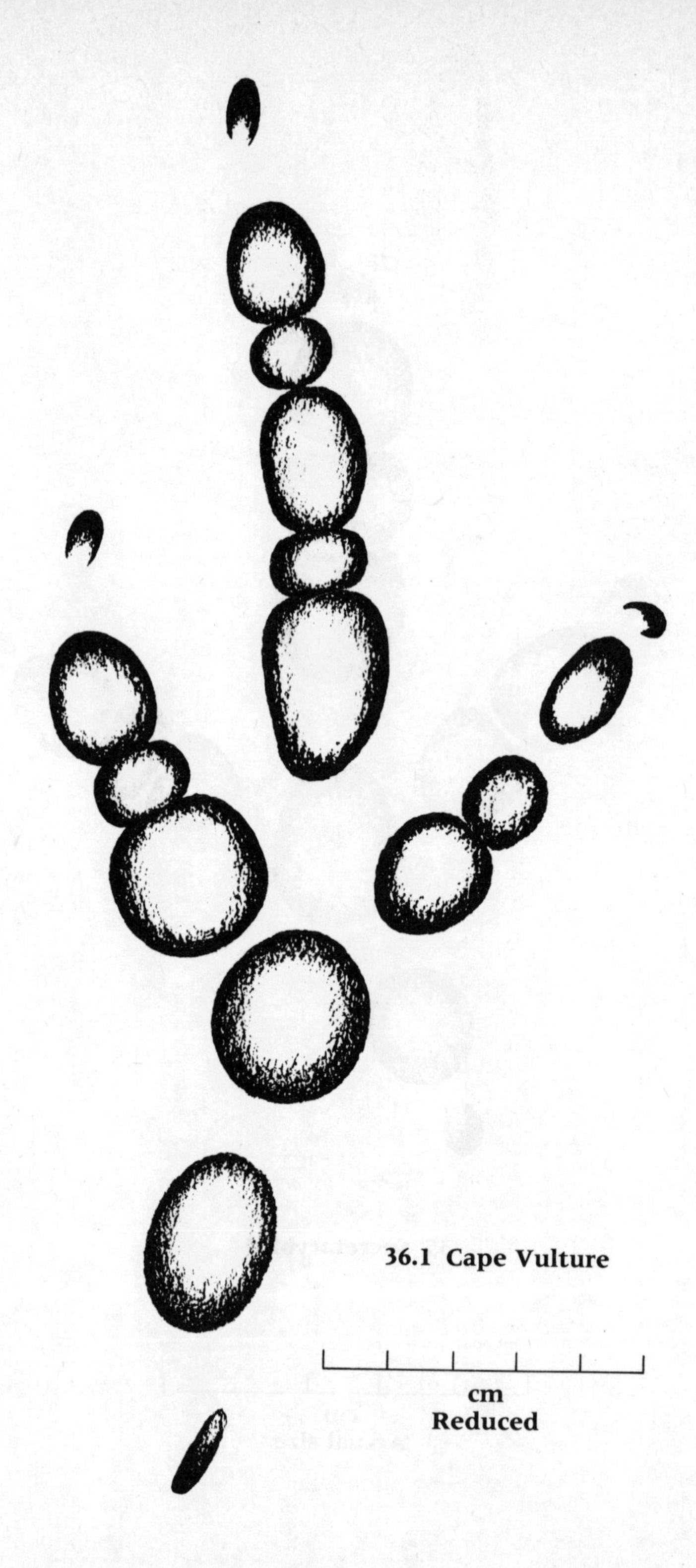

36.1 Cape Vulture

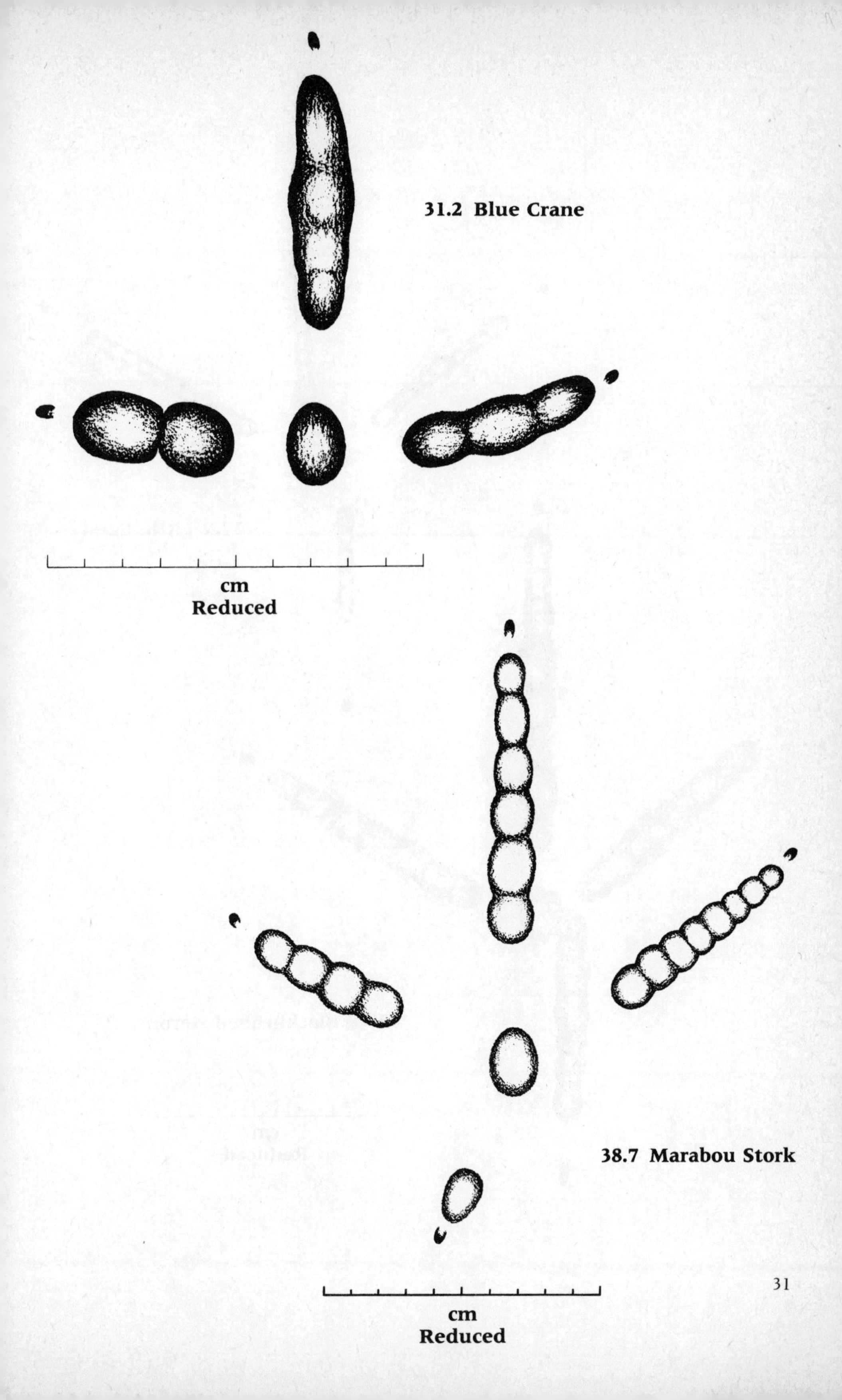

31.2 Blue Crane

38.7 Marabou Stork

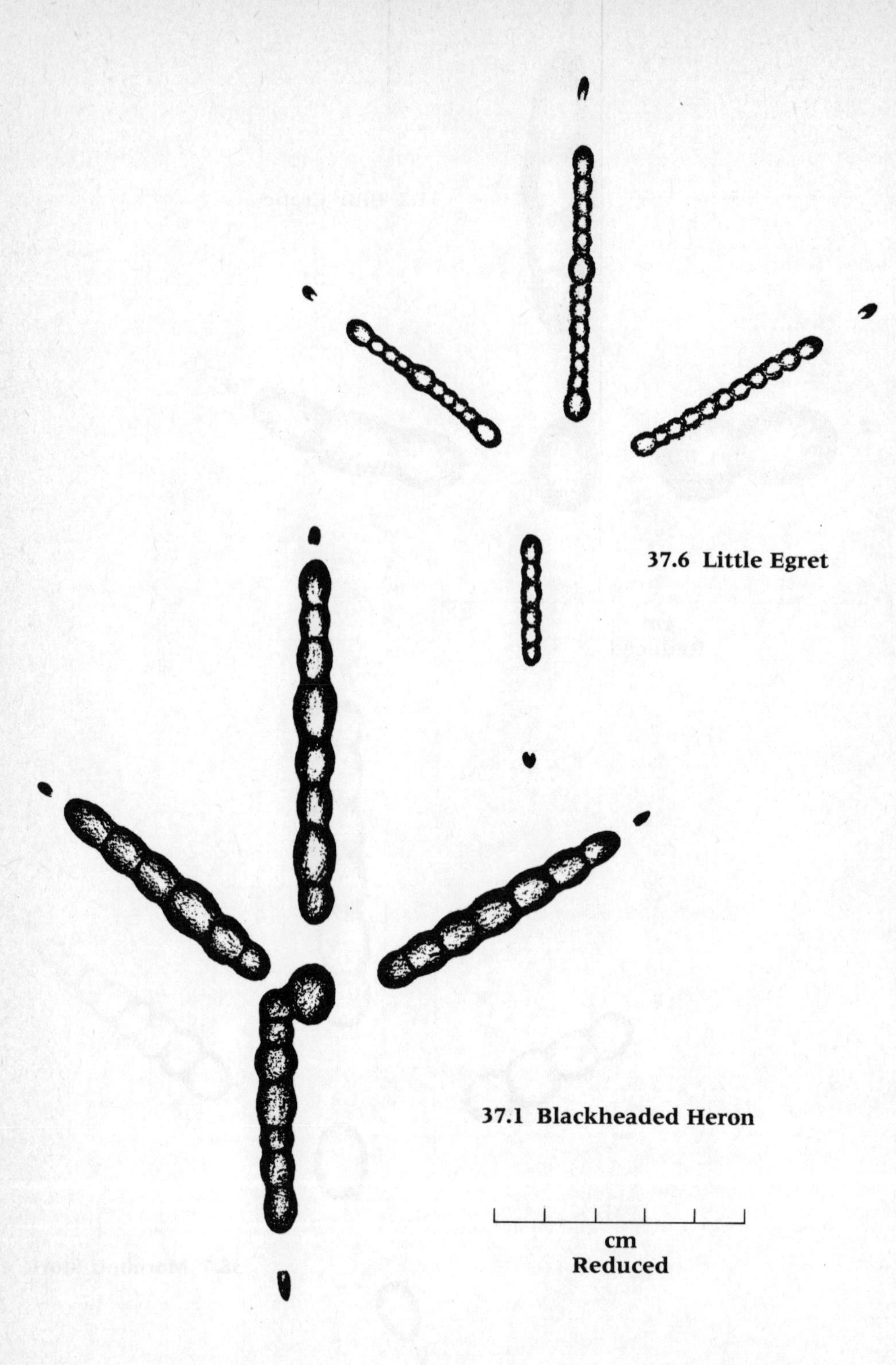

37.6 Little Egret

37.1 Blackheaded Heron

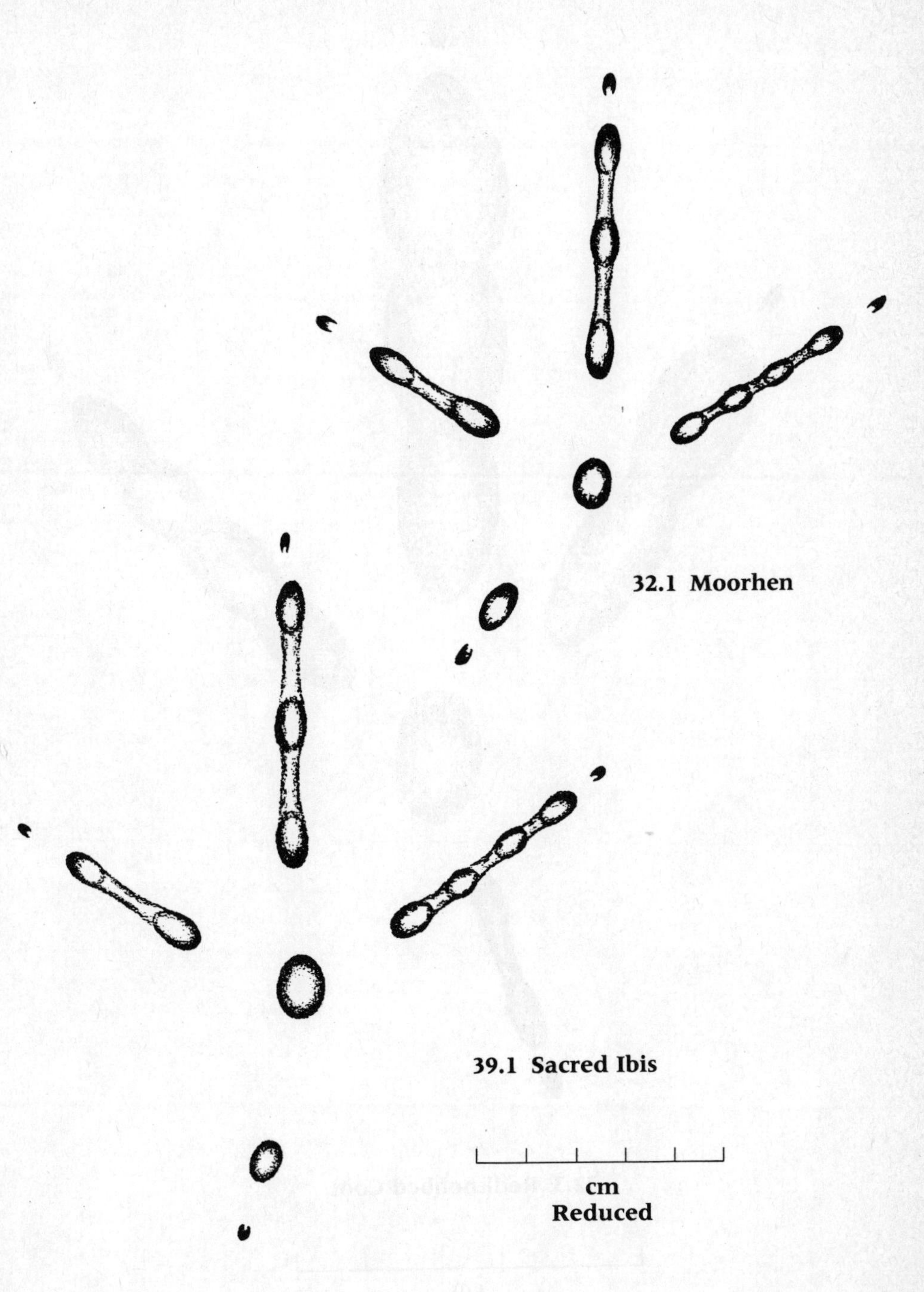

32.1 Moorhen

39.1 Sacred Ibis

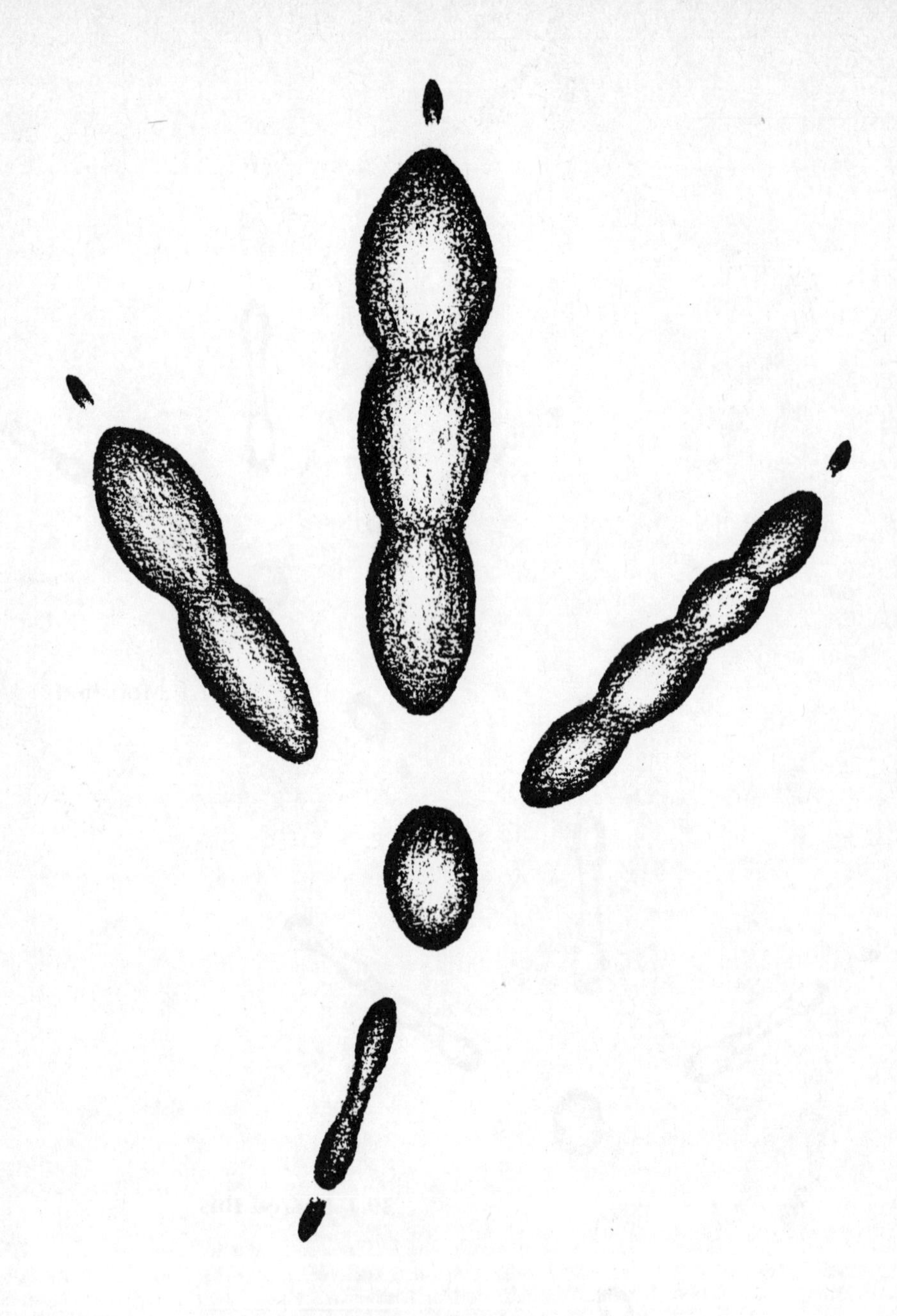

32.3 Redknobbed Coot

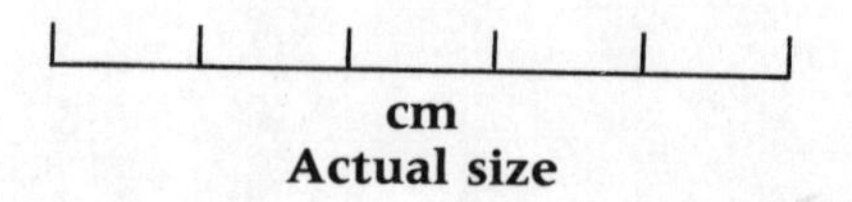

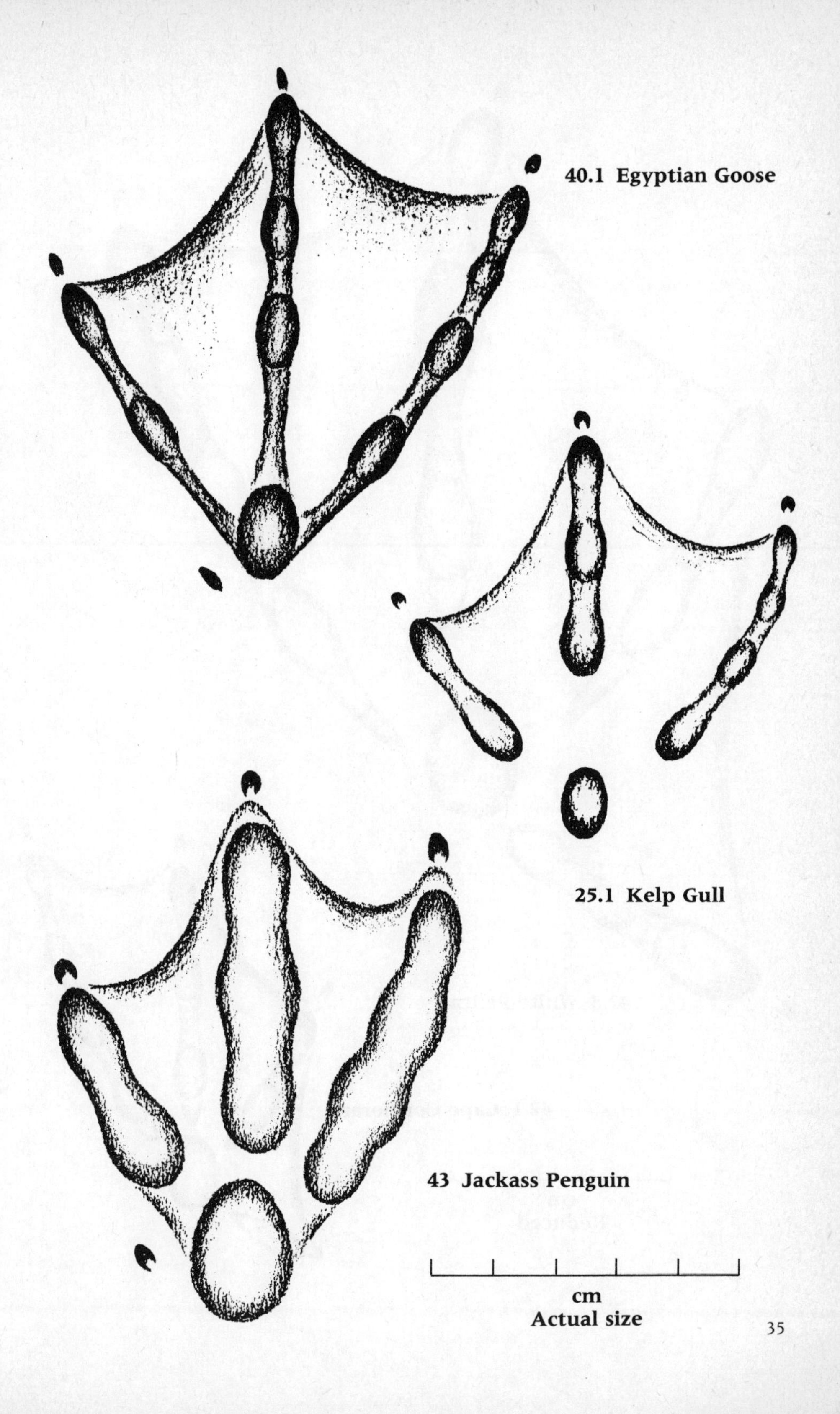
40.1 Egyptian Goose
25.1 Kelp Gull
43 Jackass Penguin
cm
Actual size

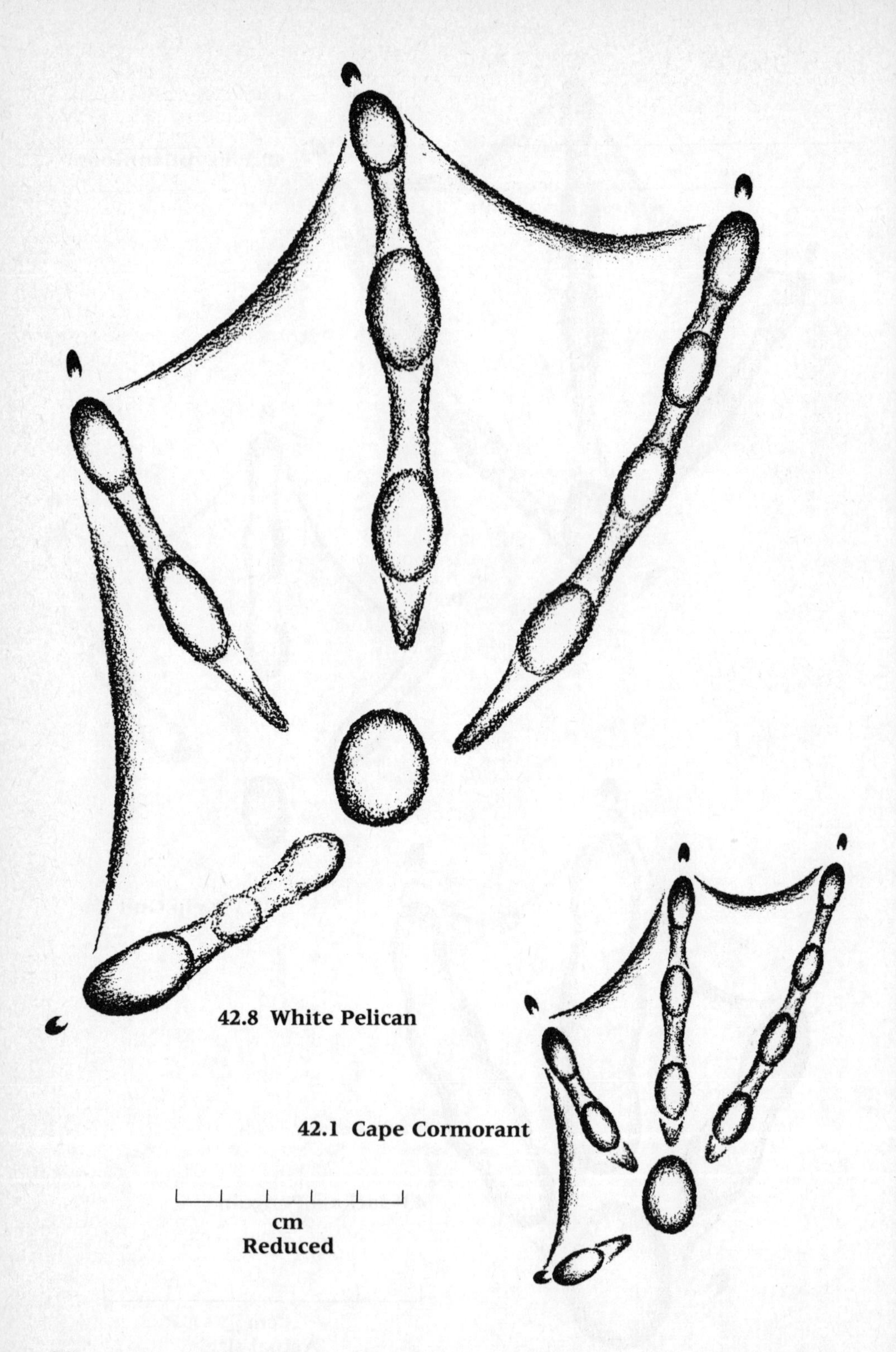

42.8 White Pelican

42.1 Cape Cormorant

Order PASSERIFORMES
Passerines
15

This order is represented by over 5 000 species of small to large birds. Most of the commoner small birds that one sees from day to day are passerines, which have also been called 'perching birds' or 'songbirds'. They are represented in southern Africa by 29 families. Most of these, about 80 per cent, are largely arboreal, but some have terrestrial habits. Their feet have three toes in front and one behind, adapted for perching. An example of the spoor of a small passerine of the family Ploceidae is shown in Fig. 15.1.

Corvus capensis
Black Crow
Swartkraai
16.1

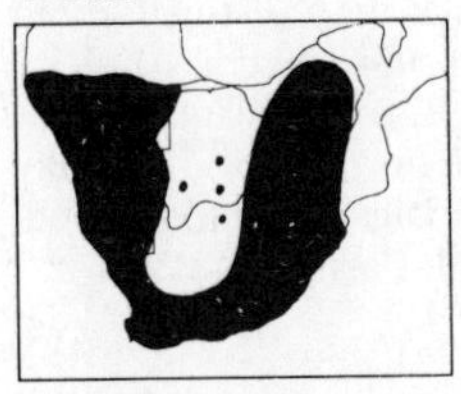

A large bird, glossy black all over, with a slender bill. TL 48–53 cm. **Habitat:** Open grassland, alpine meadows, cultivated fields, exotic plantations, *Acacia* savanna, riverine trees in desert. **Habits:** Usually in pairs with permanent territory, sometimes solitary or in flocks of up to 50 birds. Forages on ground. **Food:** Omnivorous.

Corvus albus
Pied Crow
Witborskraai
16.2

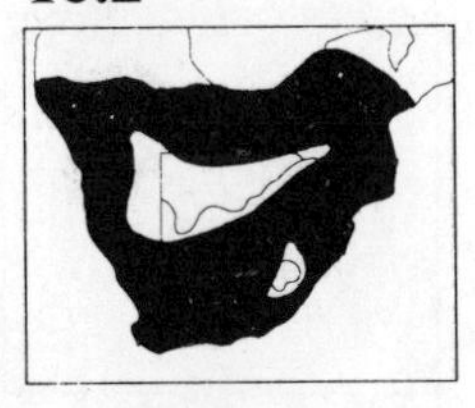

A large bird, shiny black with white breast and broad white collar on hindneck. TL 46–52 cm. **Habitat:** Savanna, farmland, urban areas, verges of roads and railways, rubbish dumps. **Habits:** Usually in pairs or small flocks, sometimes in flocks of up to 300 birds. Forages mainly on ground. **Food:** Primarily plant material, also arthropods, molluscs, frogs, reptiles, fish, birds, eggs, small mammals, carrion.

Corvus albicollis
Whitenecked Raven
Withalskraai
16.3

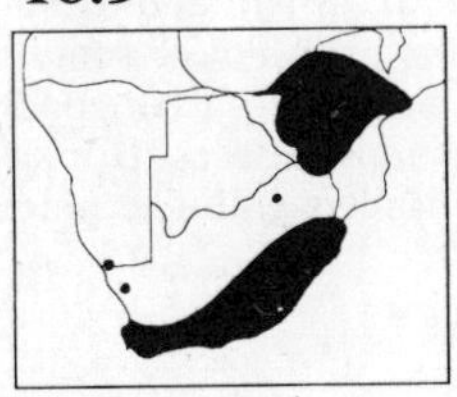

A large bird, glossy black with white collar on hindneck. Bill massive, arched, with white tip. TL 50–54 cm. **Habitat:** Mainly mountains, gorges, cliffs. **Habits:** Usually singly or in pairs, sometimes in flocks of up to 150 birds. **Food:** Carrion, insects, birds' eggs, fruit, grain, birds, mammals, reptiles.

Corvus splendens
House Crow
Huiskraai
16.4

A large bird, shiny black with grey breast, nape and mantle. TL about 43 cm. **Habitat:** Urban. **Habits:** Usually gregarious, in flocks of up to 50 birds. Forages on ground. Roosts in trees. **Food:** Omnivorous.

Tockus flavirostris
Yellowbilled Hornbill
Geelbekneushoringvoël

18.1

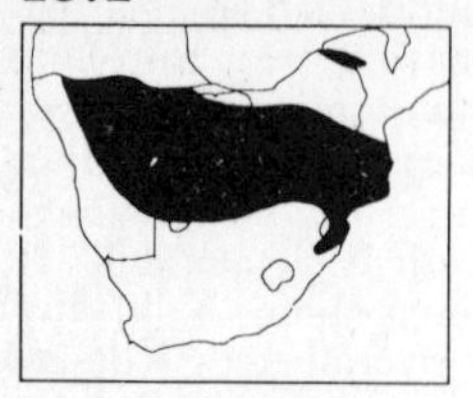

A medium-sized bird with a large, deep yellow bill and yellow eyes. Wings boldly mottled black-and-white. TL 48–60 cm. Mass ♂ 211 g, ♀ 168 g. **Habitat:** Bushveld, woodland, savanna, arid thornveld. **Habits:** Solitary, in pairs, or in small groups. Forages mostly on ground, but also in trees. Runs on ground. Similar to Redbilled Hornbill. **Food:** Rodents, insects, scorpions, solifuges, centipedes, seeds, fruit.

Bucorvus leadbeateri
Ground Hornbill
Bromvoël

18.9

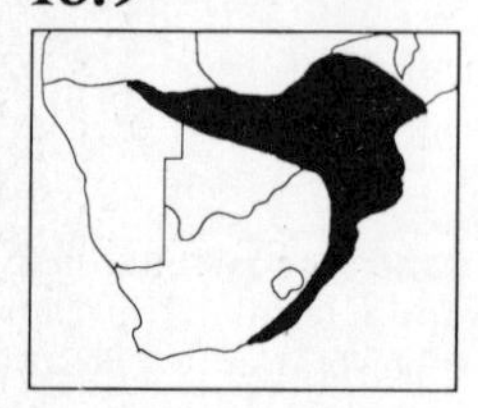

A very large turkey-like bird, mostly black with distinctive red wattles. In flight primaries are white. TL 90–129 cm. Mass ♂ 3,67 kg, ♀ 2,23–2,3 kg. **Habitat:** Any woodland, savanna, open grassveld, agricultural land. **Habits:** In pairs or groups of usually not more than 8 birds. Forages on ground. Digs with bill for food. Roosts in groups at ends of branches. **Food:** Carnivorous.

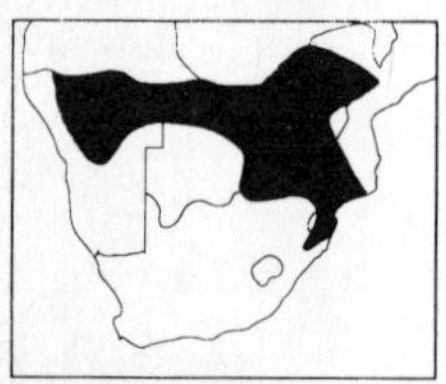

18.2 Redbilled Hornbill

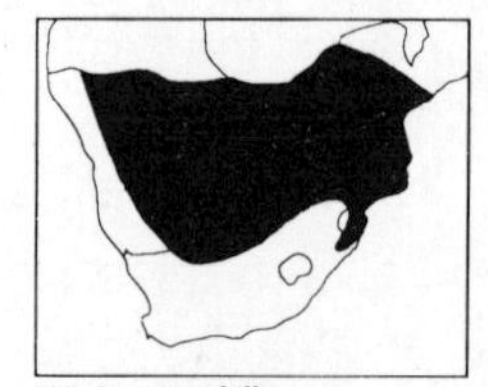

18.3 Grey Hornbill

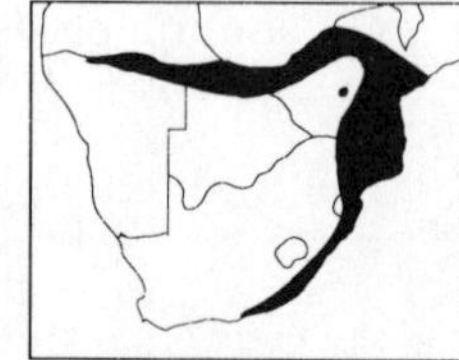

18.4 Crowned Hornbill

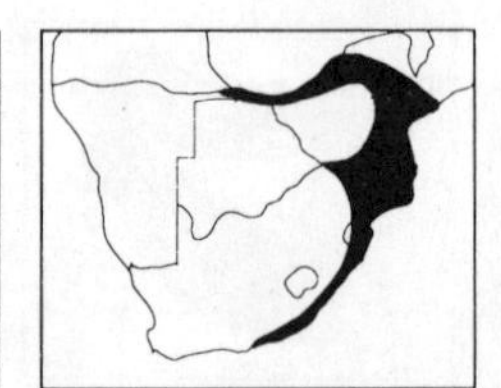

18.7 Trumpeter Hornbill

Streptopelia capicola
Cape Turtle Dove
Gewone Tortelduif

19.3

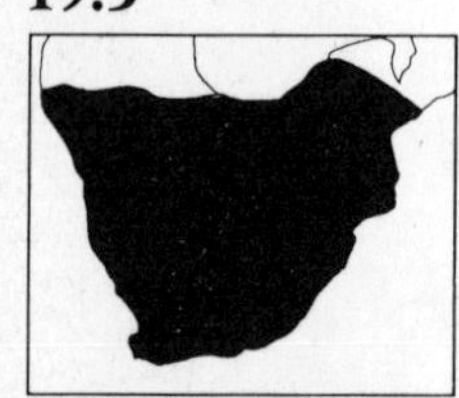

A medium-sized dove, clear grey, darker on back. Black collar on hindneck. TL 26–28 cm. Mass 153 g. **Habitat:** Woodland (not forest), savanna, riverine bush, farmland, urban and rural gardens, city parks. **Habits:** Solitary, in pairs or in flocks, sometimes of several hundred birds, especially at waterholes or good food supply. Forages on ground. Rests in tops of trees. Drinks mainly in morning. **Food:** Seeds, insects, fallen grain, termite alates.

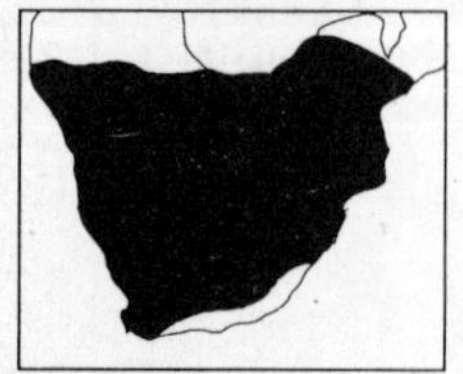

19.1 Namaqua Dove

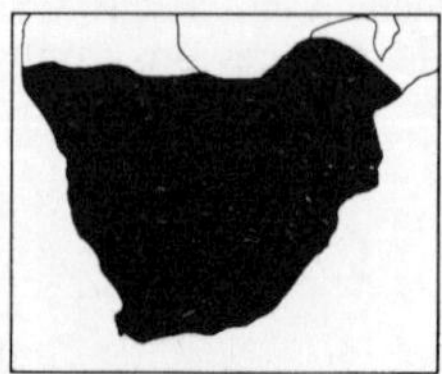

19.2 Laughing Dove

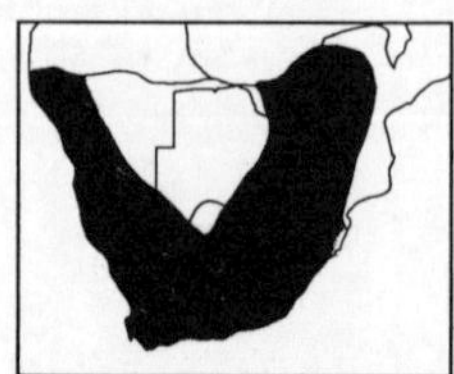

19.4 Rock Pigeon

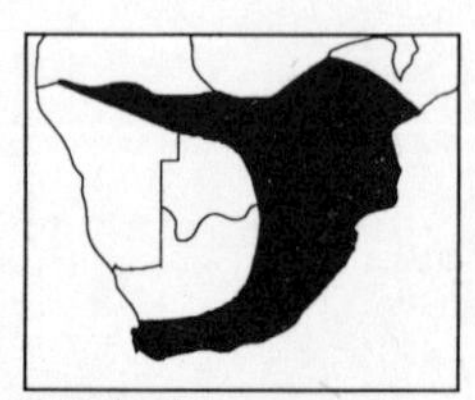

19.7 Redeyed Dove

Vanellus armatus
Blacksmith Plover
Bontkiewiet

22.2

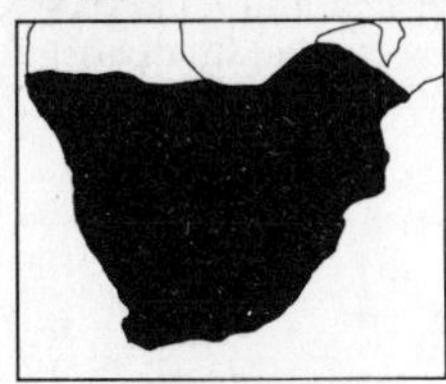

A medium-sized plover, boldly pied black-and-white with greyish back and wings. The underparts are black from chin to upper belly. TL 30 cm. Mass 157 g. **Habitat:** Shorelines of dams, pans, vleis, sewage ponds; also wet pastures, short grassy verges of inland waters, large lawns, playing fields. Less often, tidal flats in bays and lagoons. **Habits:** Often solitary or in pairs. Non-breeding birds may gather in loose flocks of 20–30, sometimes more. Silent when foraging or resting, usually calling only in flight or when alarmed. Forages in short grass or on shorelines, stepping quickly in short bursts, stopping to peck suddenly at food. **Food:** Insects, worms, molluscs.

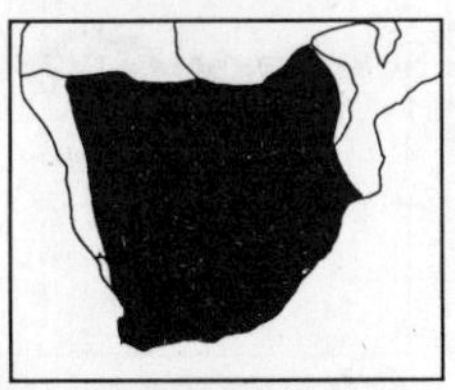

22.1 Crowned Plover

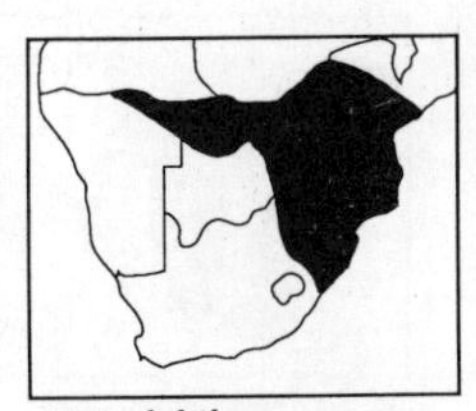

22.3 Wattled Plover

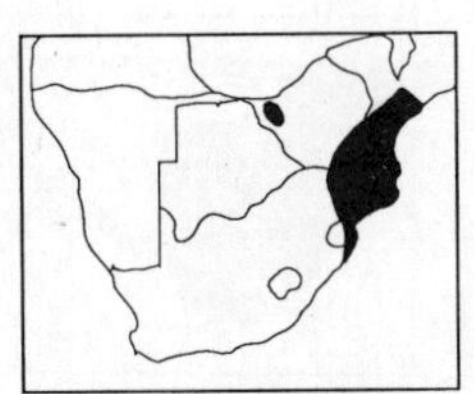

22.4 Lesser Blackwinged Plover

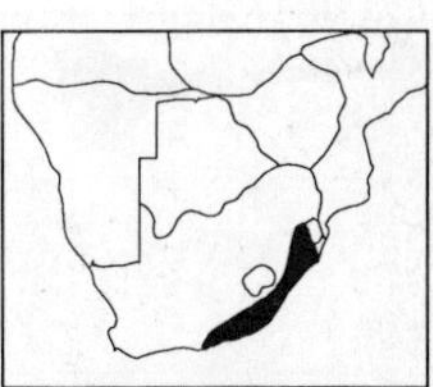

22.5 Blackwinged Plover

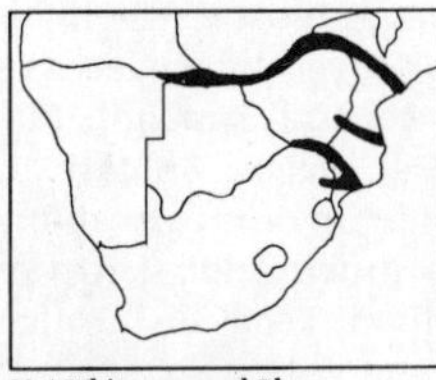

22.6 Whitecrowned Plover

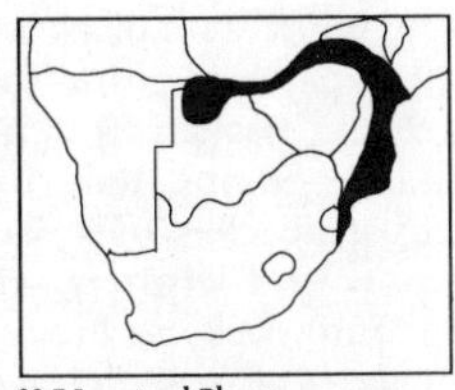

22.7 Longtoed Plover

Burhinus capensis
Spotted Dikkop (Cape Dikkop)
Dikkop

23.1

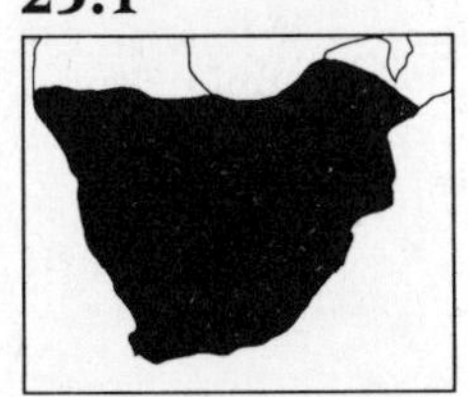

A medium-sized plover-like bird with conspicuous, large, yellow eyes and yellow legs and feet. Spotted dark brown on buff above, below white, faintly washed cinnamon and streaked brown on chest. TL 43–44 cm. Mass 450 g. **Habitat:** Open grassland near trees or bushes, savanna, stony semi-desert with scrub (less often), wide marine beaches; also agricultural land, large lawns, playing fields, airfields. **Habits:** Solitary or in pairs when breeding, otherwise may be gregarious in flocks of 40–50 birds. Mainly crepuscular and nocturnal, but also active on cloudy days. Vocal at night and on heavily overcast days, especially after rain. **Food:** Insects, crustaceans, molluscs, grass seeds, frogs.

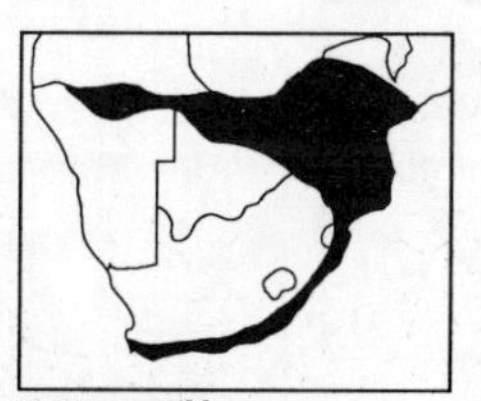

23.2 Water Dikkop

Rhinoptilus chalcopterus
Bronzewinged Courser
Bronsvlerkdrawwertjie

24.1

Medium-sized bird with long legs. Bold brown and white facial markings. Upper parts and breast dull brown, white belly, dark brown breastband, red legs. TL about 25 cm. **Habitat:** Woodland with scrub layer, *Acacia* savanna. At night moves to open grassland, roads, tracks, and clearings. **Habits:** Solitary, in pairs or small groups. Mainly nocturnal, spends day among small bushes in woodland. **Food:** Insects.

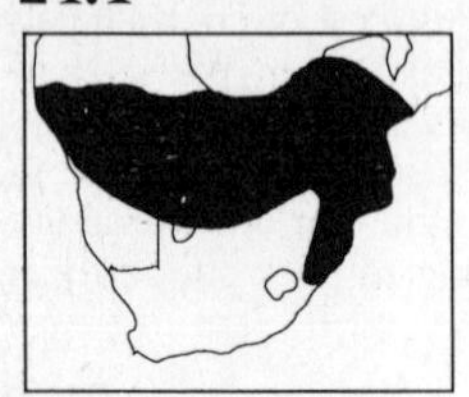

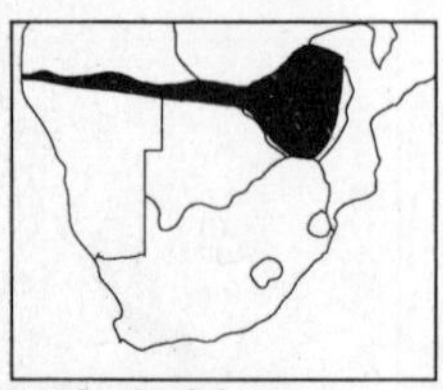
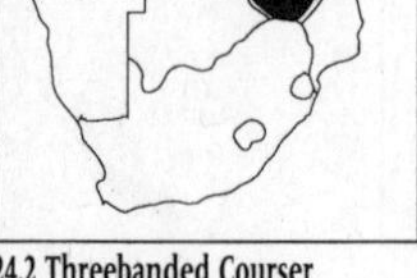

24.2 Threebanded Courser

24.3 Doublebanded Courser

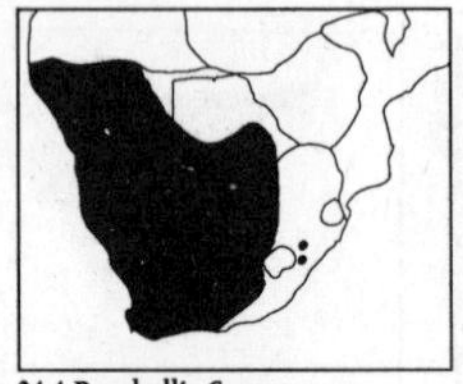

24.4 Burchell's Courser

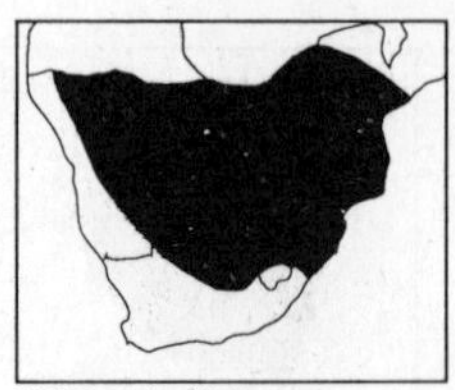

24.5 Temminck's Courser

Larus dominicanus
Kelp Gull
(Southern Blackbacked Gull)
Swartrugmeeu

25.1

A large bird, mostly white with black wings and back. The bill is bright yellow with a patch of scarlet near tip of lower jaw. TL 56–60 cm. Mass 924 g. **Habitat:** Estuaries, coastal beaches, offshore waters, rubbish dumps. Rare inland. **Habits:** Solitary or gregarious. Forages on beaches, over water or on dumps, walking or flying. Flight slow and leisurely with much gliding. Drops molluscs from air onto rocks to break them. **Food:** Fish, offal, sandmussels, limpets, insects, birds' eggs and young.

Coturnix coturnix
Common Quail
Afrikaanse Kwartel

26.1

A small bird, mottled fawn, streaked with white. TL 16–18 cm. Mass ♂ 92 g, ♀ 102 g. **Habitat:** Open grassland, lightly wooded savanna, cultivated fields. Non-breeding birds also occur in karoo, Kalahari sandveld and semi-desert. **Habits:** Usually singly or in pairs. Sits close in grass, and when flushed does not fly far, pitching suddenly into cover. Calls mainly morning and evening, as well as at night. Roosts on ground in coveys. **Food:** Seeds, buds, tubers, flowers, leaves, arthropods, worms, snails.

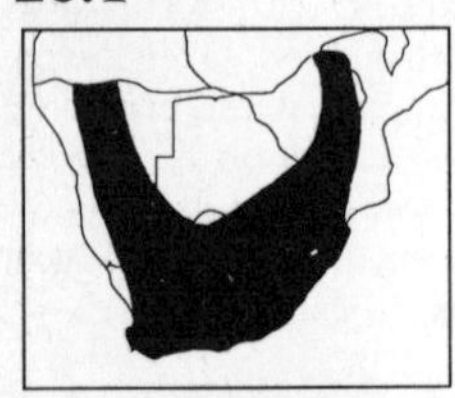

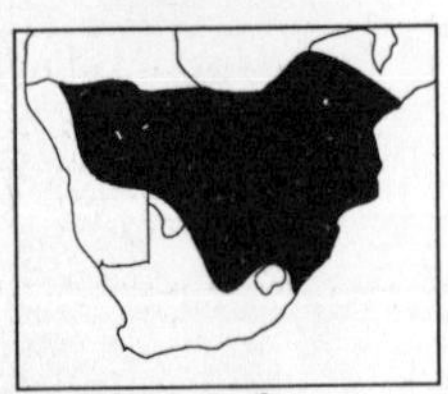

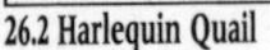

26.2 Harlequin Quail

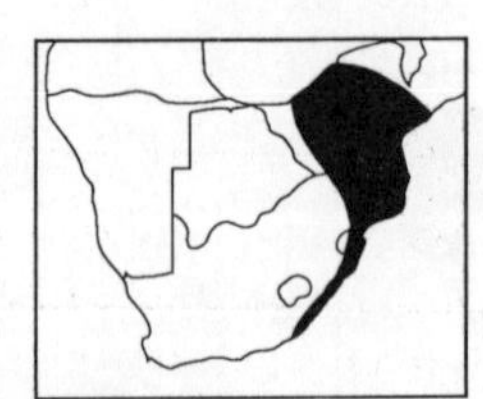

26.3 Blue Quail

Kelp Gull

Redbilled Francolin

Small Spotted Cat

Yellow Mongoose

Francolinus swainsonii
Swainson's Francolin
Bosveldfisant

27.10

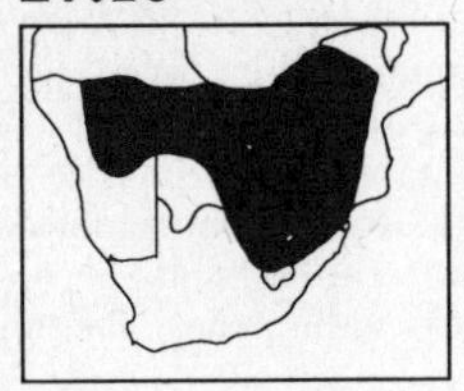

A medium-sized bird, brown above and below, streaked with black. Bill dark above, red below. Face and throat red. Legs black. TL ♂ 38 cm, ♀ 33 cm. Mass ♂ 706 g, ♀ 505 g. **Habitat:** Bushveld, edges of woodland in grass and thickets, cultivated lands, savanna, grassveld with scattered woody vegetation, riverine bush, rank vegetation around vleis. **Habits:** Solitary, in pairs or in coveys of up to 8 birds. Feeds in clearings and open fields, seeking cover in dense vegetation when disturbed. Shy and wary. Calls from tree or termite mount at dawn and dusk. Roosts in trees at night. Drinks morning and evening. **Food:** Seeds, berries, shoots, roots, bulbs, insects, molluscs.

Francolinus capensis
Cape Francolin
Kaapse Fisant

27.12

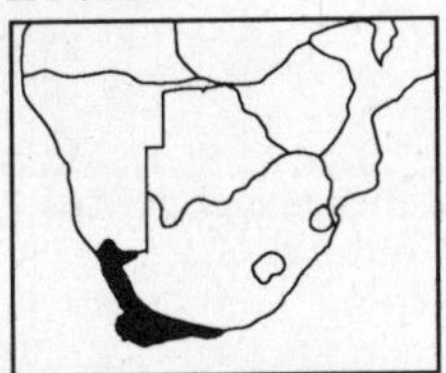

A large bird, it looks uniformly dark at a distance. The base of the bill and legs is dull orange. The belly broadly streaked white. TL 40–42 cm. Mass ♂ 600–915 g, ♀ 435–659 g. **Habitat:** Dense riverine scrub in drier areas, coastal and montane fynbos, exotic *Acacia* thickets. **Habits:** Usually pairs or coveys of up to 20 birds. Very noisy morning and evening. Often tame, feeding in clearings at edge of bush and in farmyards. Flies reluctantly but well. Prefers to run into cover. May land in trees when flushed. Roosts in trees. **Food:** Seeds, shoots, leaves, bulbs, corms, berries, insects, molluscs.

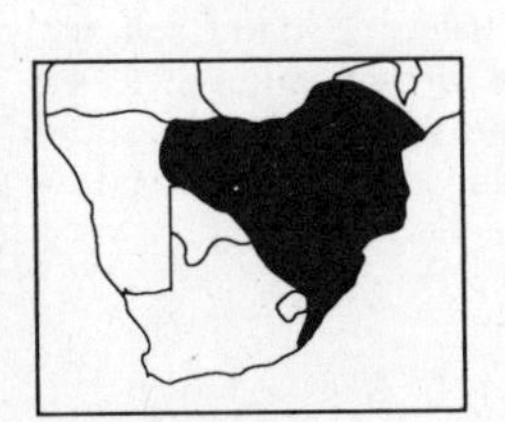

27.1 Coqui Francolin

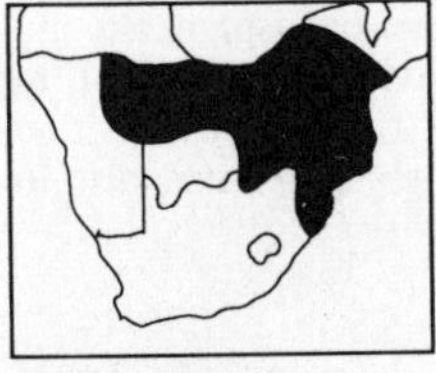

27.2 Crested Francolin

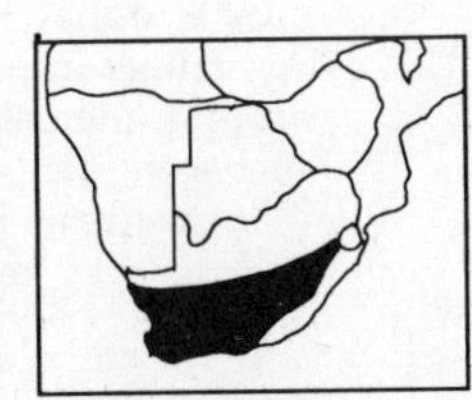

27.3 Greywing Francolin

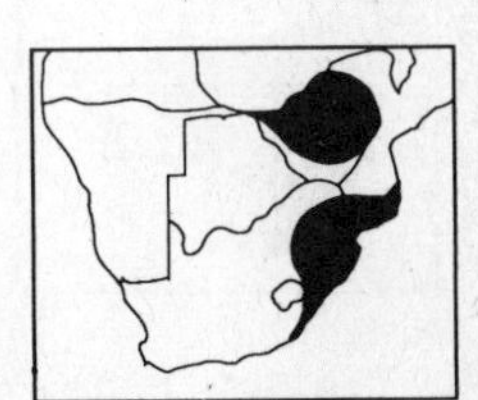

27.4 Shelley's Francolin

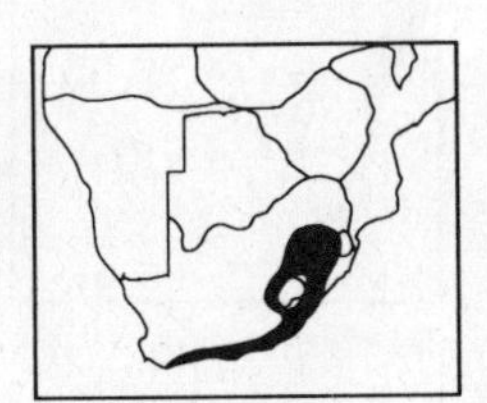

27.5 Redwing Francolin

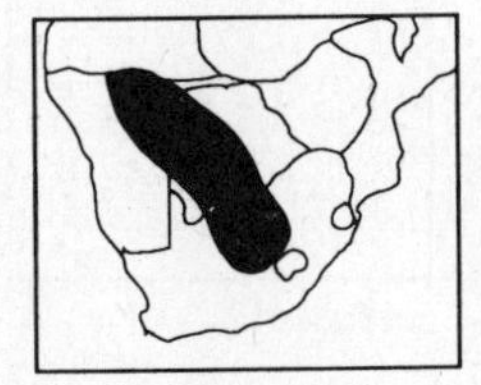

27.6 Orange River Francolin

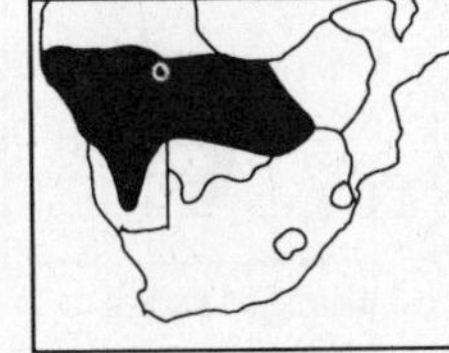

27.8 Redbilled Francolin

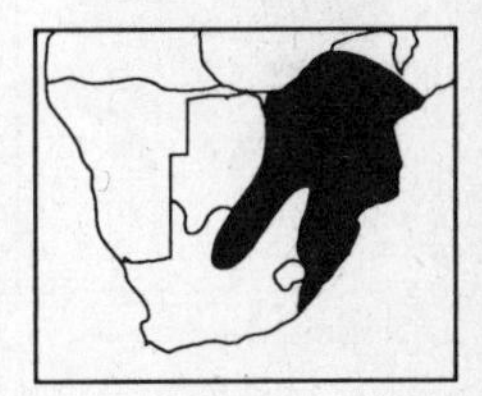

27.9 Natal Francolin

Numida meleagris
Helmeted Guineafowl
Gewone Tarentaal

28.1

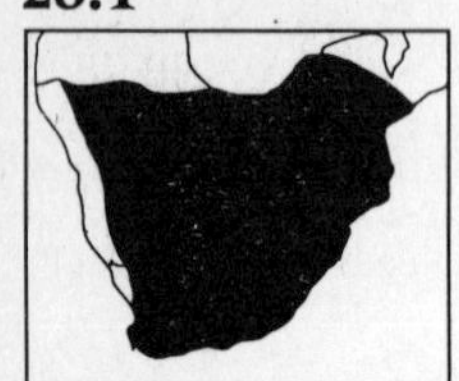

About the size of a domestic chicken, slate grey, finely spotted white. Head blue and red with conspicuous horny casque on top. TL 53–58 cm. Mass 1,35 kg. **Habitat:** Open grassland, vleis, savanna, cultivated lands, edge of karoo scrub, bushveld. **Habits:** Highly gregarious, especially when not breeding, flocks may number several hundred birds. Usually in pairs when breeding. Forages in flocks in open ground, scratching for food with feet or bill. Runs fast when disturbed. Flies well, taking to trees when hard pressed, uttering cackling alarm notes. Roosts communally in trees at night. Walks in single file to waterhole. **Food:** Seeds, bulbs, tubers, berries, insects, snails, ticks, millipedes, fallen grain.

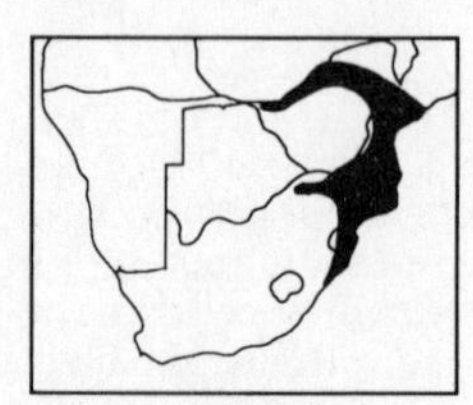

28.2 Crested Guineafowl

Eupodotis afra
Black Korhaan
Swartkorhaan

30.1

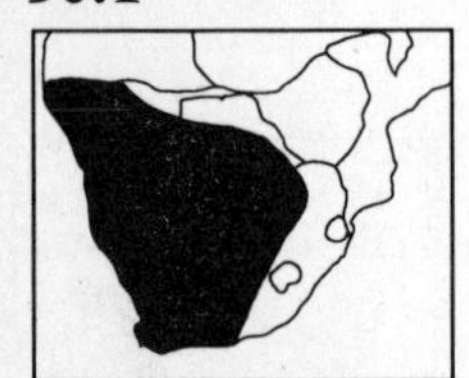

A medium-sized bird with yellow legs. Male: head, neck and belly black, with large white patches on sides of head. Back finely barred buff-and-black. Female: inconspicuous and secretive. Head, neck and breast buff, paler on breast. Belly black. TL 50–53 cm. Mass ♂ 716 g, ♀ 670 g. **Habitat:** Open dry grassland, karoo, Kalahari sandveld, arid scrub, semi-desert, fallow lands, coastal dunes in southwestern Cape. **Habits:** Usually solitary, sometimes in pairs. In display, the male flies up calling, cruises around, then slowly descends with rapidly flapping wings and increasing tempo of call. Runs with head down. **Food:** Mainly plant material, including seeds; also insects.

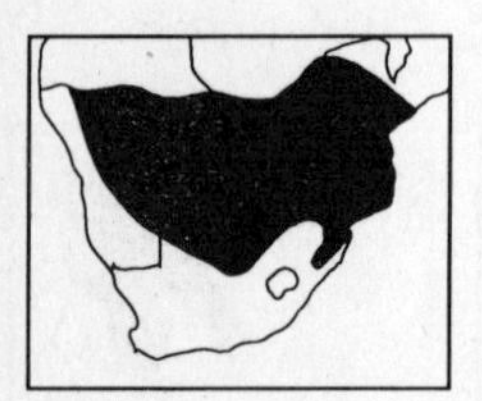

30.2 Redcrested Korhaan

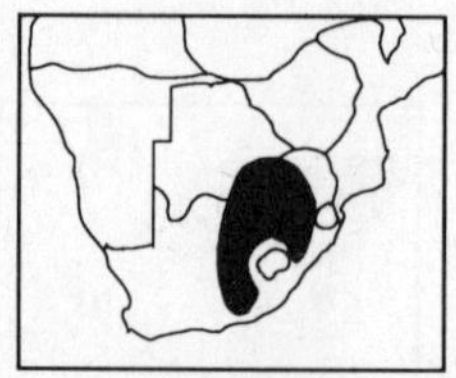

30.3 Whitebellied Korhaan

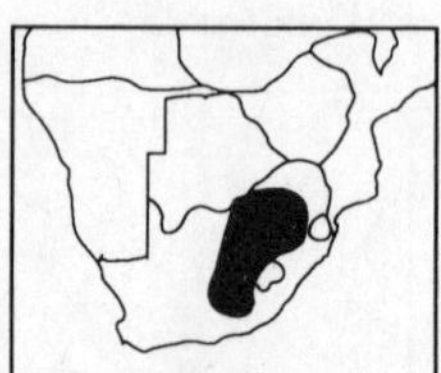

30.4 Blue Korhaan

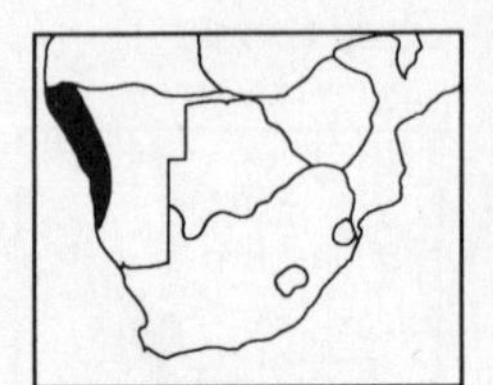

30.5 Rüppell's Korhaan

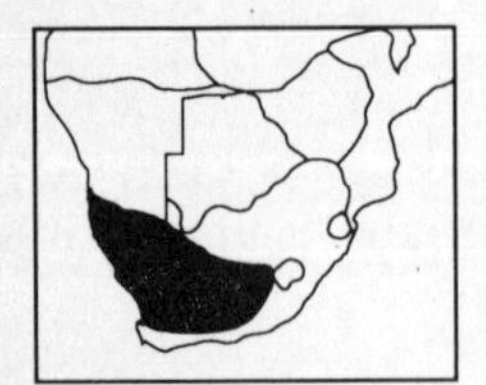

30.6 Karoo Korhaan

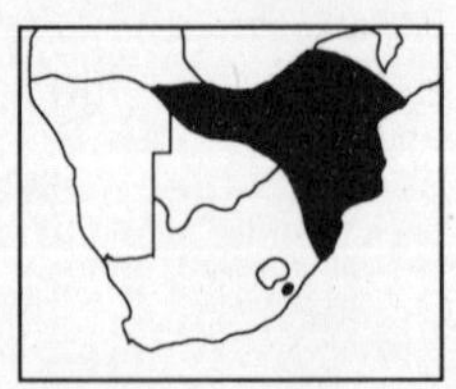

30.7 Blackbellied Korhaan

Neotis ludwigii
Ludwig's Bustard
Ludwigse Pou
30.8

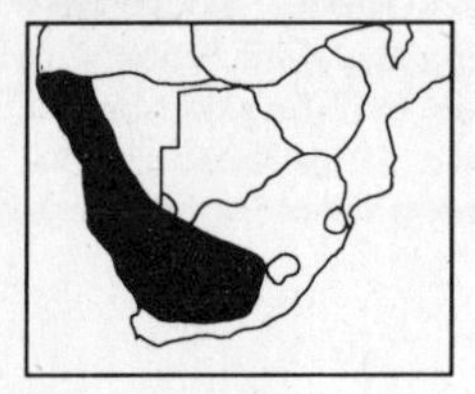

A large bird, very similar to Stanley's Bustard, but slightly smaller. Crown, face and fore-neck sooty brown, hind-neck bright chestnut. Wing coverts brown-and-white, belly white. TL ♂ 78–95 cm, ♀ 76–85 cm. Mass 3,4 kg. **Habitat:** Dry open plains, from grassland to desert. **Habits:** Solitary or in groups of up to 6 birds. Shy and wary. Flies readily when disturbed, usually until out of sight. Sometimes squats to avoid detection. Male displays with inflated neck, raised tail and fanned undertail coverts. **Food:** Insects, seeds, small vertebrates, plant material.

Neotis denhami
Stanley's Bustard
Veldpou
30.9

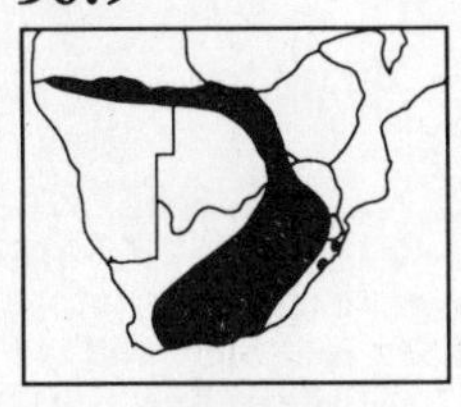

A large bird, brown above with deep rich rufous hind-neck and mantle, fore-neck grey and white below. Black-and-white area on wing. Crown black with narrow white median streak. TL ♂ 100–110 cm, ♀ 80–87 cm. Mass ♂ 9 kg, ♀ 4 kg. **Habitat:** Montane and highland grassveld, savanna, karoo scrub. **Habits:** Solitary or in pairs when breeding, otherwise in groups of up to 10 birds. Shy and wary, walking away quickly when disturbed. Flies strongly, usually at some height. In courtship male inflates neck, raises tail and fans undertail coverts. **Food:** Mainly insects; also millipedes, grass, seeds, flowers, lizards, small rodents.

Ardeotis kori
Kori Bustard
Gompou
30.10

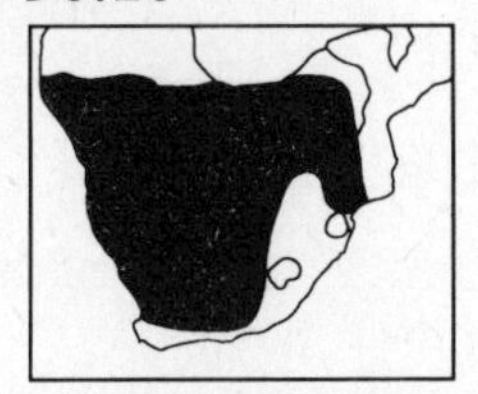

A very large bird, greyish brown above, white belly, neck and breast finely barred (looks grey at a distance), head slightly crested and with longish bill. Walks with bill angled slightly upwards. TL ♂ 1,2–1,5 m, ♀ 1,05–1,2 m. Mass ♂ 13,5–19 kg. **Habitat:** Open plains of karoo, highveld grassland, Kalahari sandveld, arid scrub, Namib Desert, lightly wooded savanna, bushveld. **Habits:** Solitary or in pairs when breeding, otherwise gregarious in flocks of up to 40 or more birds. Walks slowly and sedately when foraging. When disturbed walks quickly. Flies reluctantly, but powerfully. **Food:** Insects, small vertebrates, seeds, carrion, gum.

Anthropoides paradisea
Blue Crane
Bloukraanvoël

31.2

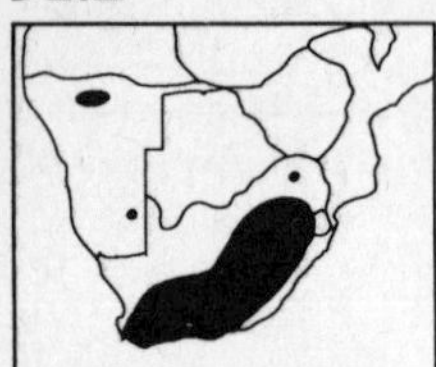

A very large bird with long legs, plain blue-grey all over, with long slate-grey tertails curving to the ground. TL 1 m. **Spoor:** Fig. 31.2. **Habitat:** Midland and highland grassveld, edge of karoo, cultivated land, edges of vleis. **Habits:** Highly gregarious when not breeding, otherwise in pairs or family groups. Flocks usually 30–40 birds, sometimes up to 300. Flies strongly and soars well. Roosts on ground or in shallow water. Wary when breeding, otherwise fairly tame. Often performs display dances in groups or pairs. **Food:** Frogs, reptiles, insects, fish, grain, green shoots, grass seeds.

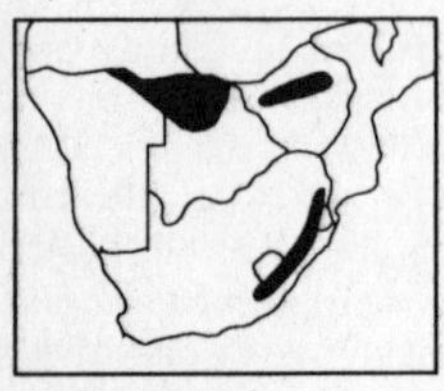

31.1 Wattled Crane

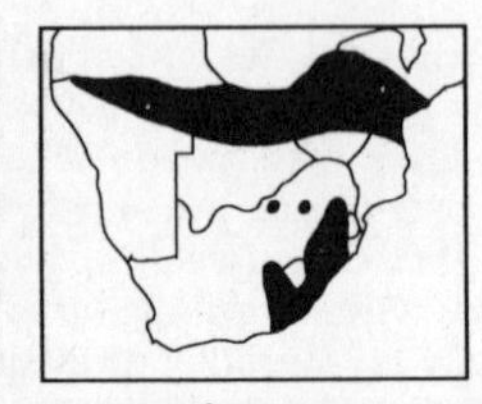

31.3 Crowned Crane

Gallinula chloropus
Moorhen
Waterhoender

32.1

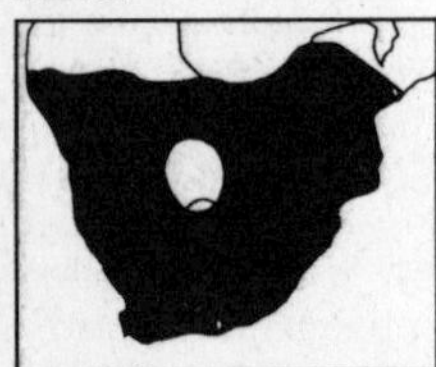

A medium-sized bird, slaty black all over, except for white undertail and white streaks on flanks. It has a red shield and bill with yellow tip. TL 30–36 cm. Mass 247 g. **Habitat:** Reedbeds, marshes, marginal vegetation of lakes, rivers, pans and sewage ponds. **Habits:** Solitary or in small family groups. Spends most of the day swimming in green water, wading in shallows or walking over nearby wet grasslands. Roosts in reeds or low bushes. **Food:** Water plants, seeds, berries, molluscs, worms, arachnids, insects, tadpoles, offal, carrion.

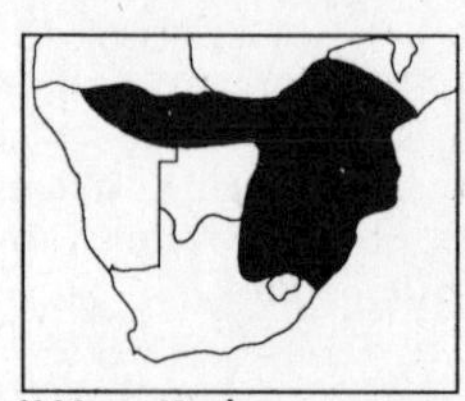

32.2 Lesser Moorhen

Fulica cristata
Redknobbed Coot
Bleshoender

32.3

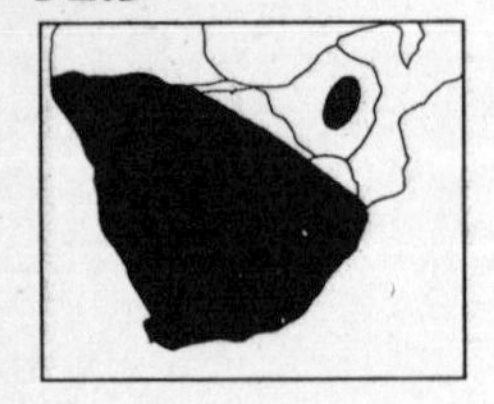

A medium-sized bird, all black with white bill and frontal shield, backed by two dark red knobs. TL 43 cm. Mass 737 g. **Habitat:** Almost all inland waters, especially with floating water plants. Less commonly on rivers and coastal lagoons. **Habits:** Usually in pairs or large flocks of over 1 000 birds. Highly gregarious when not breeding. Spends most of time swimming in open water. Forages in water from surface or by diving. Also grazes on shoreline, running to water when disturbed. **Food:** Mainly water plants and grass; also insects and seeds.

Struthio camelus
Ostrich
Volstruis

34

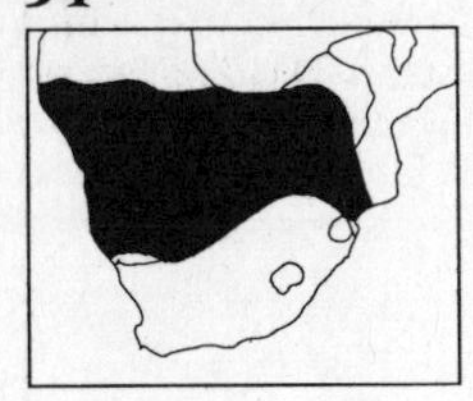

The largest living bird. Males are black with white wings, females brownish grey. TL ♂ up to 2 m. Mass 68 kg. **Habitat:** Bushveld to desert. **Habits:** Occurs in flocks of 30–40 birds when not breeding. In desert regions up to 600 birds may gather at waterholes. Can run at speeds of 50–60 km/hour. Males may perform elaborate displays in courtship and distraction when breeding. **Food:** Grass, berries, seeds, succulent plants, small reptiles, insects.

Sagittarius serpentarius
Secretarybird
Sekretarisvoël

35

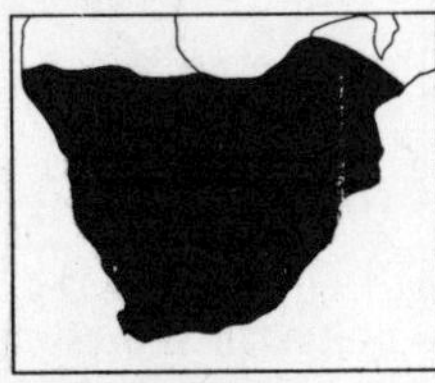

A very large bird, about 1,3 m tall, with long legs and tail. Body mainly pale grey, with belly, tibial feathering, rump and crest feathers black. The long crest is erectile. TL 1,25–1,5 m. Mass 3,9 kg. Short toes, webbed at the base, with strong claws (Fig. 35). **Habitat:** Semi-desert, grassland, savanna, open woodland, farmland, mountain slopes. **Habits:** Usually in pairs, sometimes in groups of 3 or 4. Strides slowly across veld. Catches prey on ground with bill or sometimes by stamping on it. Flies seldom but well, taking off with a run. Roosts at night on top of a bush or tree, usually a pair together. **Food:** Insects, rodents, amphibians, lizards, snakes, young hares, birds' young and eggs.

Gyps coprotheres
Cape Vulture
Kransaasvoël

36.1

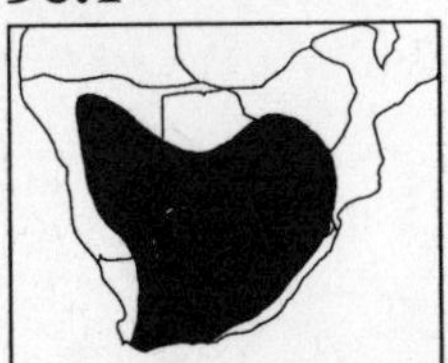

A very large vulture, pale whitish to buffy with strongly contrasting blackish wings and tail. Paired blue bare patches on either side of crop. TL 101–120 cm. Mass 8,6 kg. **Habitat:** Mostly mountainous country, or open country with isolated hills or mountains and escarpments. Less common in savanna or desert. **Habits:** Highly gregarious. Roosts and nests on precipitous cliffs. Some 2–3 hours after sunrise it soars out to forage over a wide area, often far away from mountains. Aggressive at carcasses. **Food:** Carrion and bone fragments.

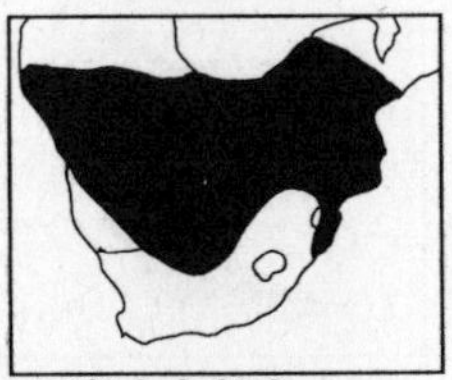

36.2 Whitebacked Vulture

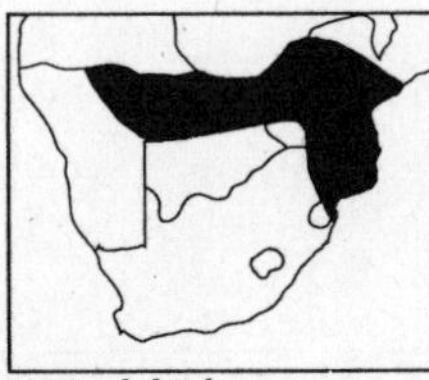

36.3 Hooded Vulture

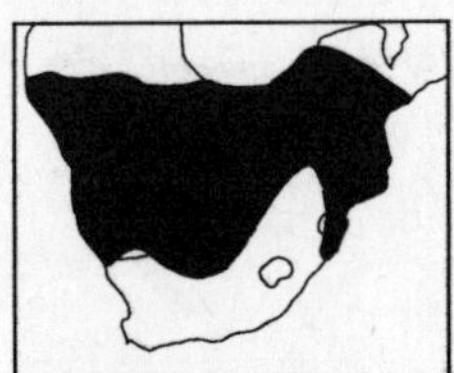

36.4 Lappetfaced Vulture

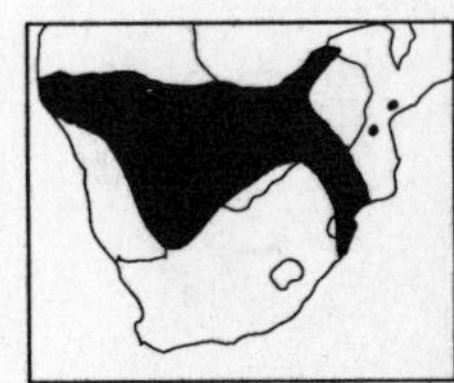

36.5 Whiteheaded Vulture

Ardea melanocephala
Blackheaded Heron
Swartkopreier

37.1

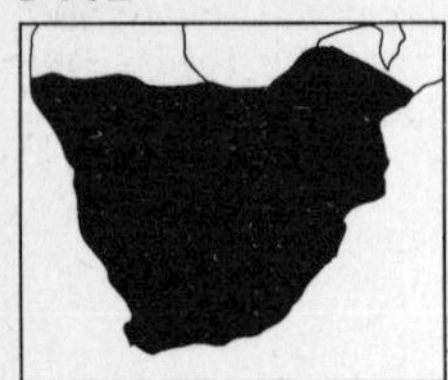

A fairly large, slender bird. Head black above, white below. The long neck is black on the hind-neck, white on the fore-neck. The underwing is white in front, black behind. TL 97 cm. Mass 1,1 kg. **Habitat:** Open grassland, fallow field, edges of inland waters, forest clearings. **Habits:** Solitary when feeding, either standing and waiting for prey, or stalking slowly. Roosts colonially in trees, reedbeds and on islands up to 30 km from feeding grounds. **Food:** Frogs, fish, crabs, insects, rodents, small birds, small reptiles, worms, spiders.

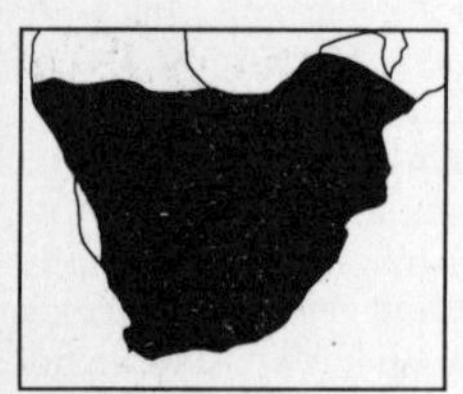

37.2 Grey Heron

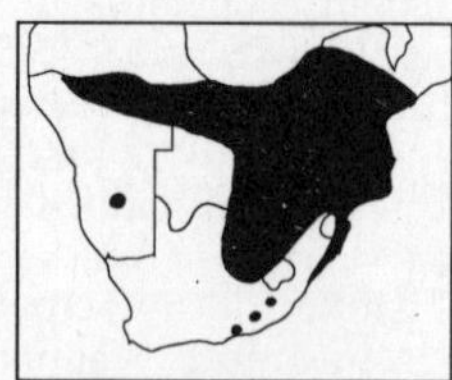

37.3 Goliath Heron

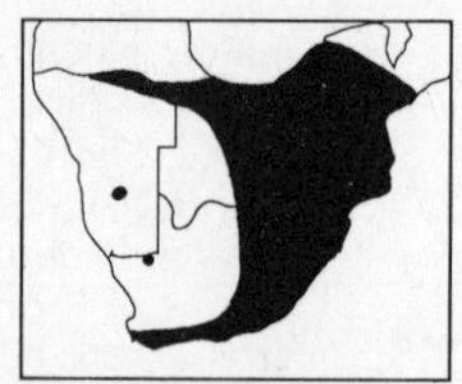

37.4 Purple Heron

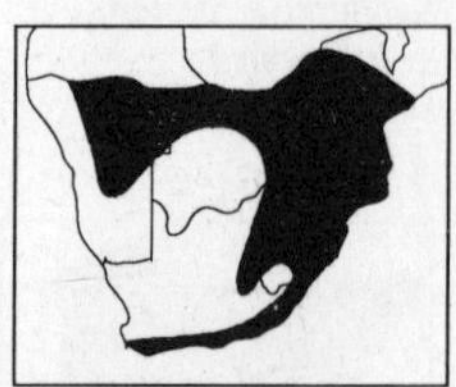

37.5 Great White Egret

Egretta garzetta
Little Egret
Kleinwitreier

37.6

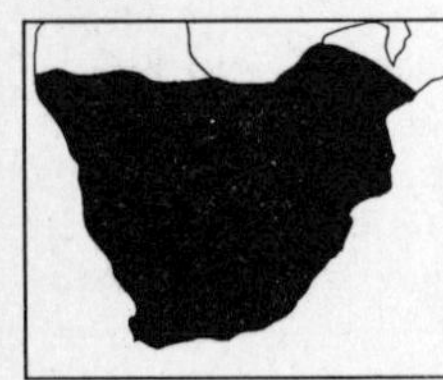

A smallish, slender white bird with a slim black bill and black legs with yellow feet. TL 64 cm. Mass 280–614 g. **Habitat:** Shores of inland and marine waters. **Habits:** Usually solitary when feeding, but may gather in hundreds at a good food supply. Roosts gregariously. Active hunter, sometimes stands and waits for prey. May fish cooperatively in groups. **Food:** Fish; also frogs, insects, crustaceans, molluscs, small lizards.

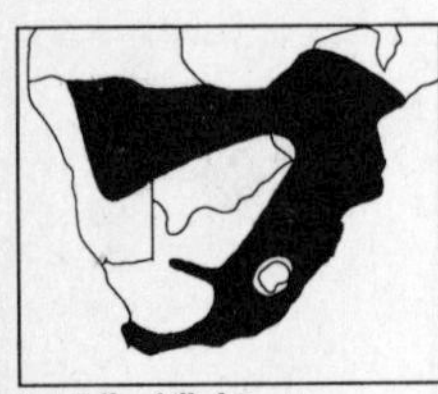

37.7 Yellowbilled Egret

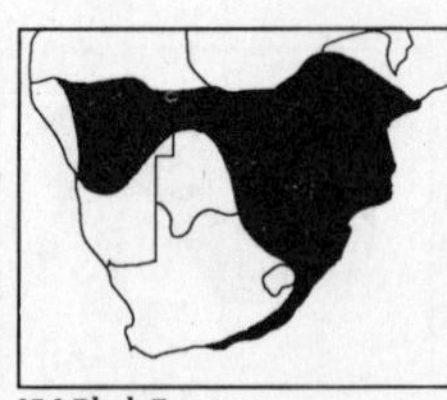

37.8 Black Egret

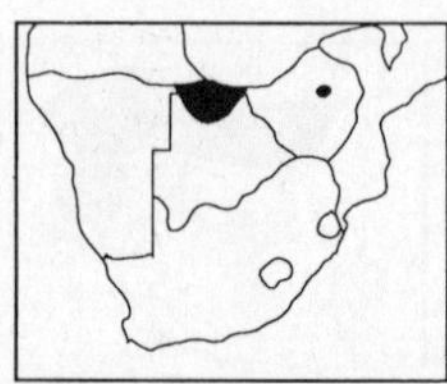

37.9 Slaty Egret

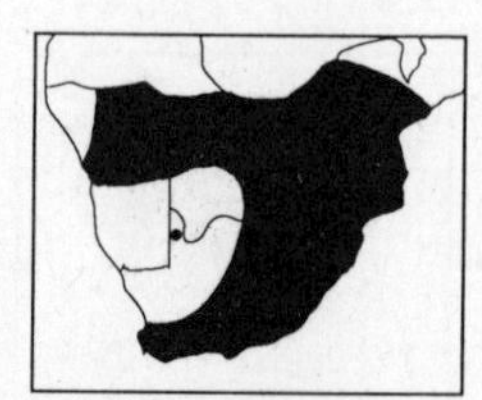

37.10 Cattle Egret

Leptoptilos crumeniferus
Marabou Stork
Maraboe

38.7

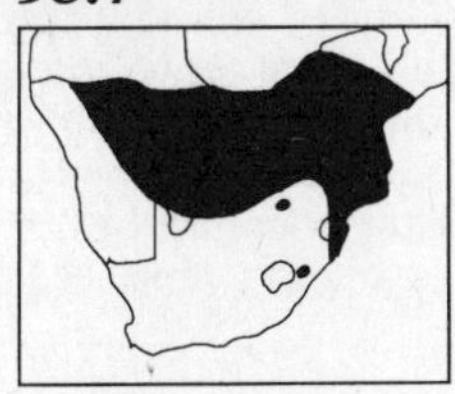

A huge bird, blackish above, white below with head naked, pinkish. The very large bill is dirty greyish, and the long legs are white. TL 1,5 m. **Habitat:** Open to semi-arid woodland, bushveld, fishing villages, rubbish tips, lake shores. **Habits:** Usually gregarious, especially around carcasses of large mammals, often with vultures. Spends most of the day standing still or squatting. Mainly a scavenger, but may forage in grassveld for insects or wade in shallow water. Takes off running with lowered head. **Food:** Carrion, refuse, rodents, insects, birds, fish, young crocodiles, lizards, snakes, frogs.

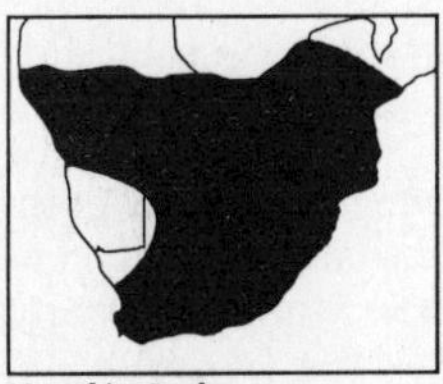

38.1 White Stork

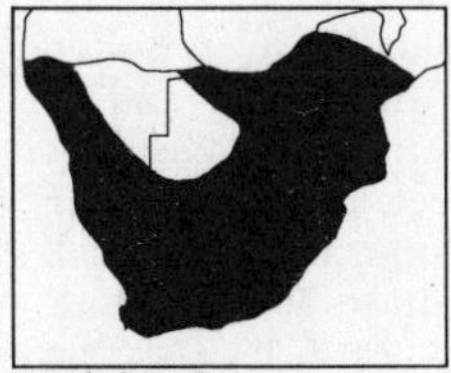

38.2 Black Stork

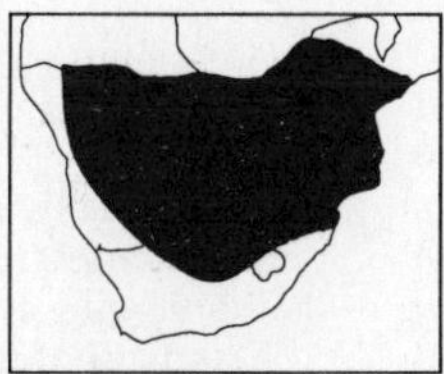

38.3 Abdim's Stork

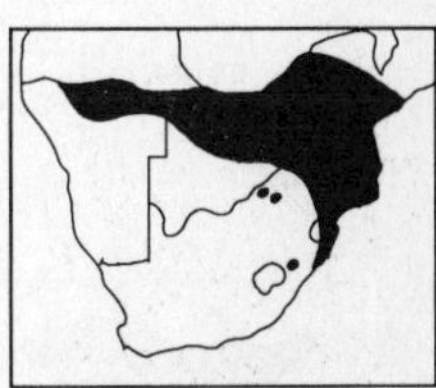

38.6 Saddlebilled Stork

Threskiornis aethiopicus
Sacred Ibis
Skoorsteenveër

39.1

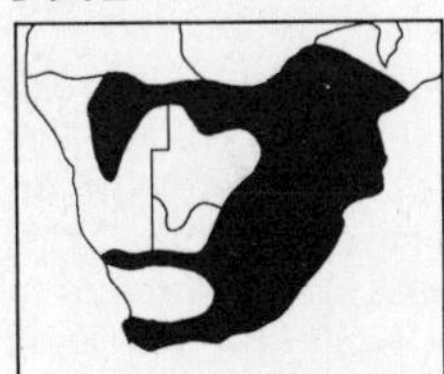

A large bird, mainly white with black head, neck and plume-like feathers on back. The bill is long and decurved. TL about 90 cm. Mass 1,3 kg. **Habitat:** Very varied, including inland waters, cultivated lands, sewage works, open grassveld, rubbish dumps, coastal lagoons, tidal flats, offshore islands. **Habits:** Gregarious, flocks may number hundreds of birds. Scavenges and forages. Roosts in trees, reedbeds or on islands. Flies in V-formation. **Food:** Very varied. Arthropods, small mammals, nestling birds, eggs, molluscs, frogs, small reptiles, carrion, seeds.

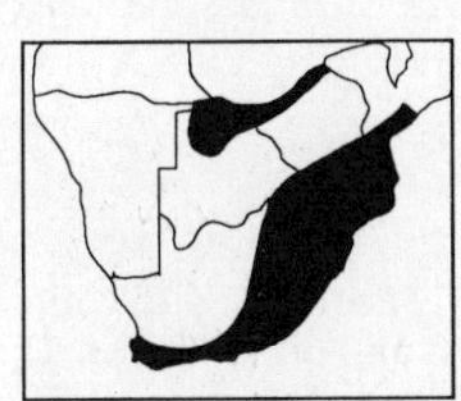

39.2 Hadeda Ibis

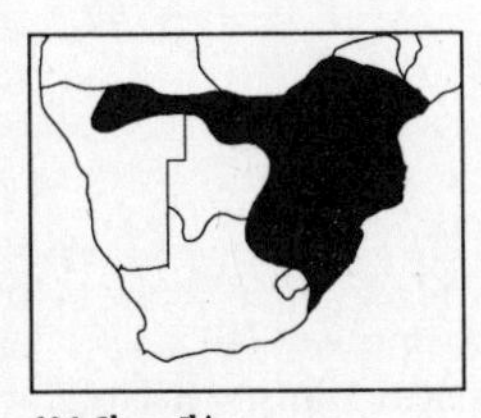

39.3 Glossy Ibis

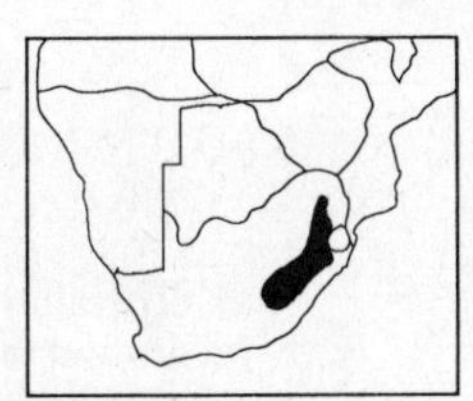

39.4 Bald Ibis

Alopochen aegyptiacus
Egyptian Goose
Kolgans

40.1

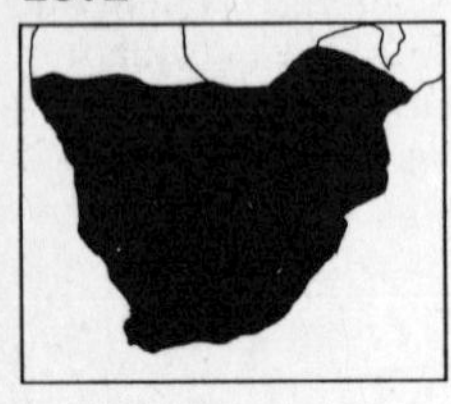

A large goose, brown above, greyish below, with dark brown patches around the eyes and on the centre of the breast, and a dark brown collar on the neck. In flight the wings are white with black primaries and green trailing edge. TL 63–73 cm. Mass ♂ 2,4 kg, ♀ 1,9 kg. **Habitat:** Most inland waters, such as rivers, dams floodplains, pans, marshes. Also estuaries, coastal lakes, cultivated fields. **Habits:** Spends much of day on shoreline or sandbank. Flies early morning and evening to grasslands and farmlands to graze, returning to water to roost on shoreline or in trees by day and after nightfall. **Food:** Grass, leaves, seeds, grain, crop seedlings, aquatic rhizomes, tubers.

Phalacrocorax capensis
Cape Cormorant
Trekduiker

42.1

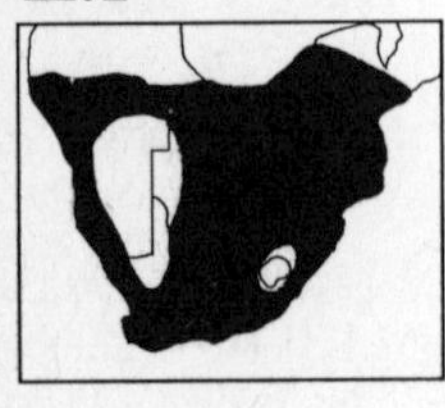

A medium-sized black bird with a long and slender bill and short tail. Face yellow, and bill, legs and feet black. TL 61–64 cm. Mass ♂ 1,3 kg, ♀ 1,2 kg. **Habitat:** Coastal waters and brackish estuaries. **Habits:** Highly gregarious. Flies in long waving lines low over the sea. Settles in large flocks to feed, diving from surface and submerging for up to 30 seconds. Roosts in large numbers on islands and cliffs. **Food:** Fish, as well as crustaceans, mussels, molluscs.

Pelecanus onocrotalus
White Pelican
Witpelikaan

42.8

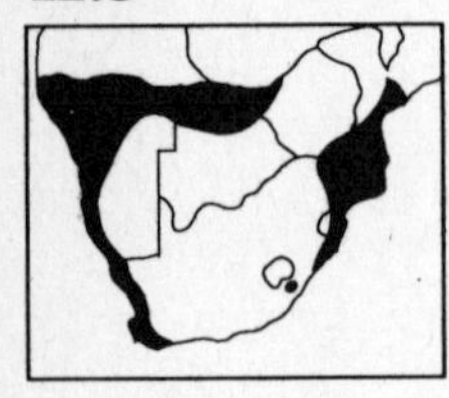

A huge white bird with a very long, straight bill and long, broad wings. Legs are short and stout, pink or yellow in colour. The large feet have long toes, all four joined by webs. TL 1,4–1,8 m. Mass ♂ 11,5 kg, ♀ 7,6 kg. **Habitat:** Coastal bays, estuaries, lakes, larger pans and dams. **Habits:** Solitary or gregarious. Forages in coordinated groups. Flies in V-formation. **Food:** Fish and some crustaceans.

Spheniscus demersus
Jackass Penguin
Brilpikkewyn

43

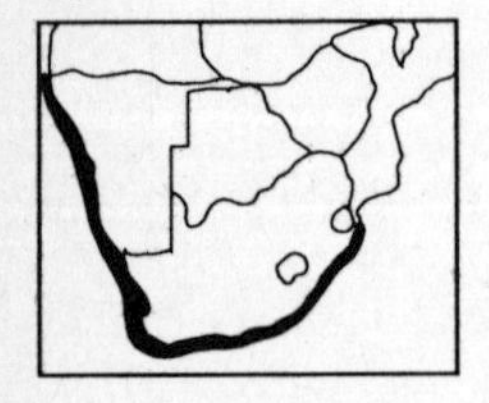

A small penguin, black above, white below. Face black with white band around face. Black band across chest and down flanks. TL about 60 cm. **Habitat:** Marine. **Habits:** Forages underwater at sea. Roosts and breeds on land. Walks with waddling gait on land. **Food:** Fish, squid.

MAMMALS

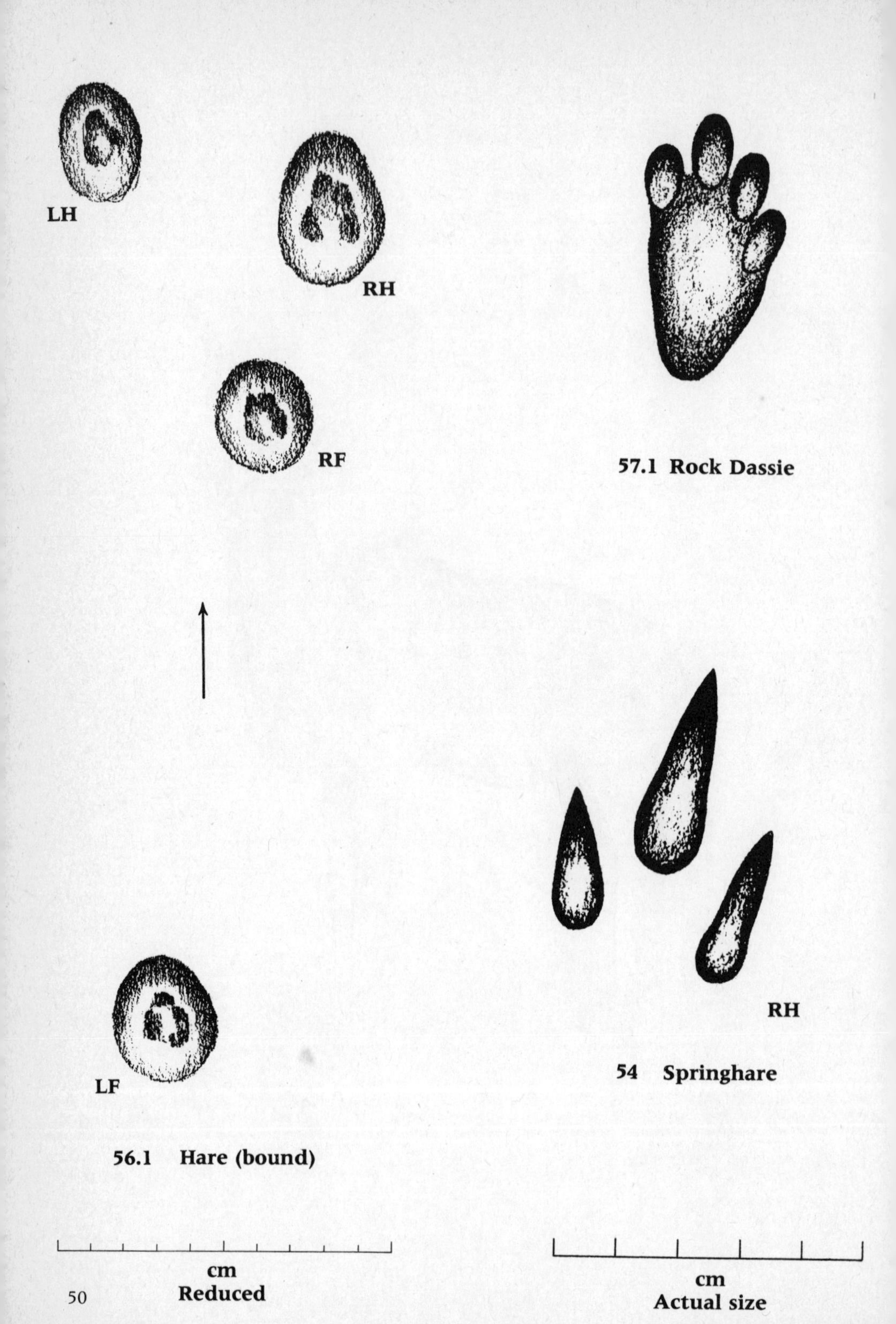

57.1 Rock Dassie

54 Springhare

56.1 Hare (bound)

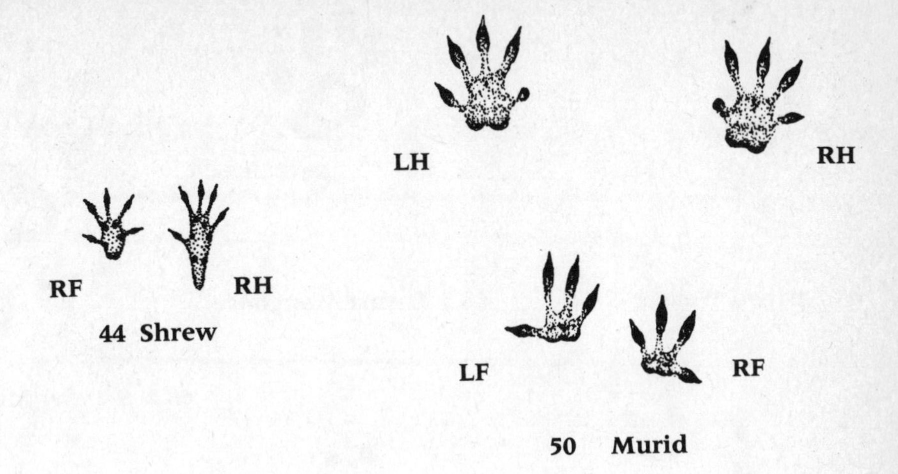

44 Shrew

50 Murid

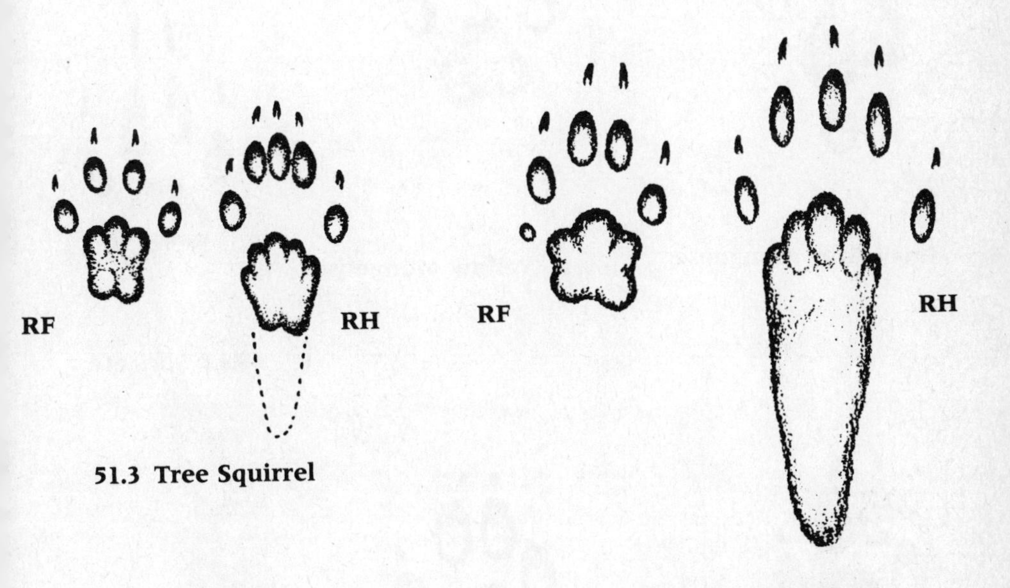

51.3 Tree Squirrel

51.1 Ground Squirrel

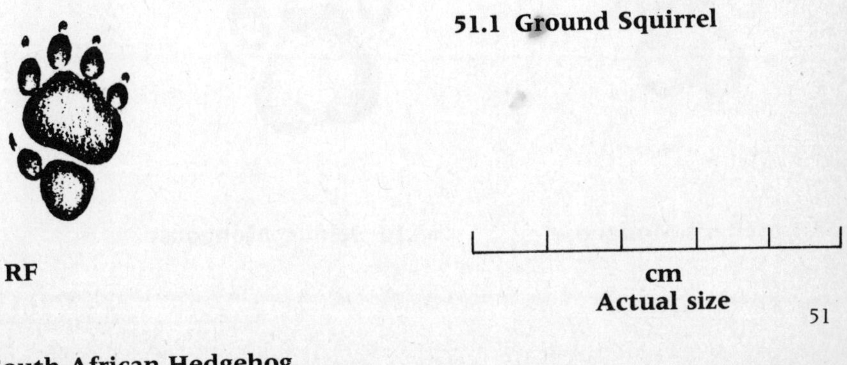

Actual size

47 South African Hedgehog

63 Striped Weasel

68.5 Dwarf Mongoose

68.2 Slender Mongoose

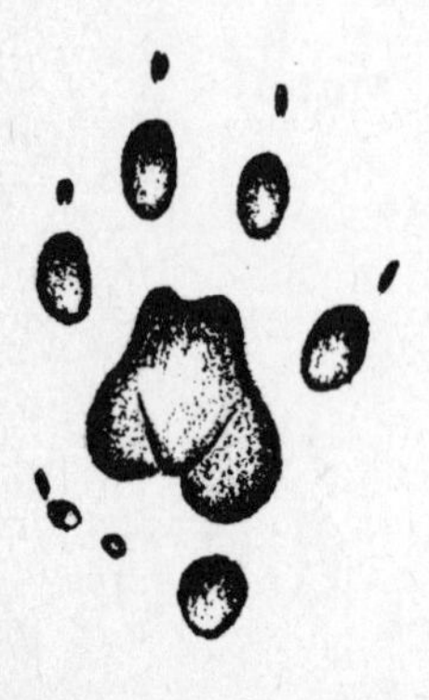

68.3 Small Grey Mongoose

68.11 Yellow Mongoose

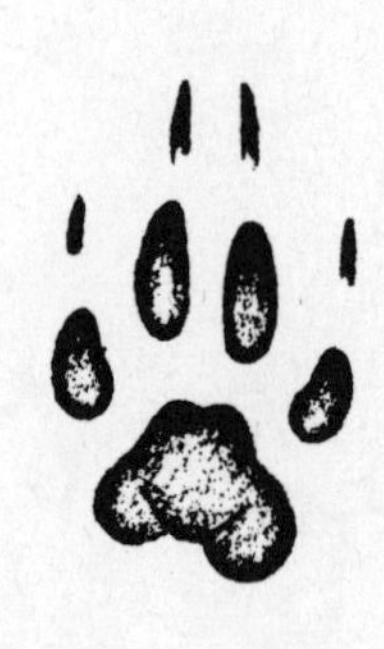

68.12 Suricate

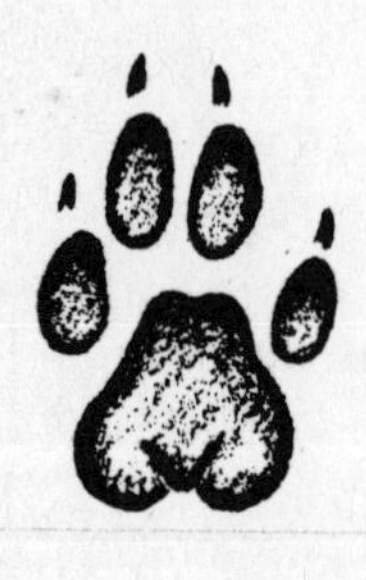

68.8 Meller's Mongoose

68.10 Selous' Mongoose

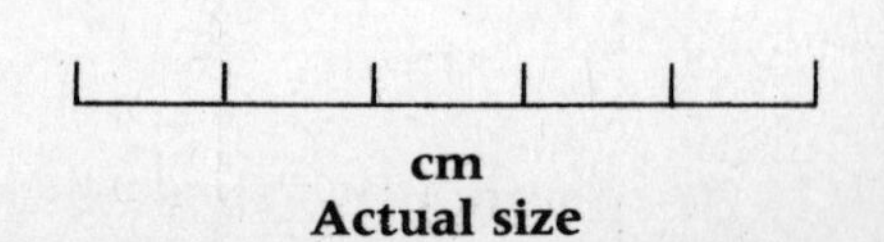

cm
Actual size

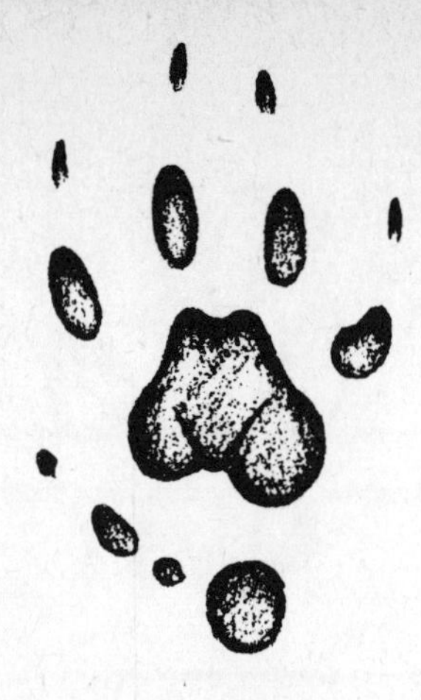

68.6 Banded Mongoose

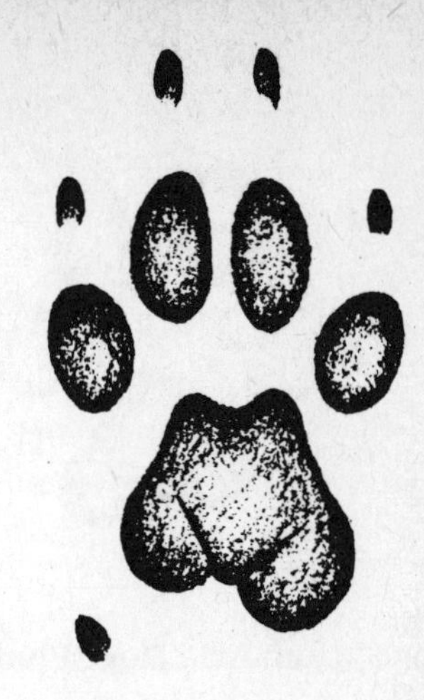

68.7 White-tailed Mongoose

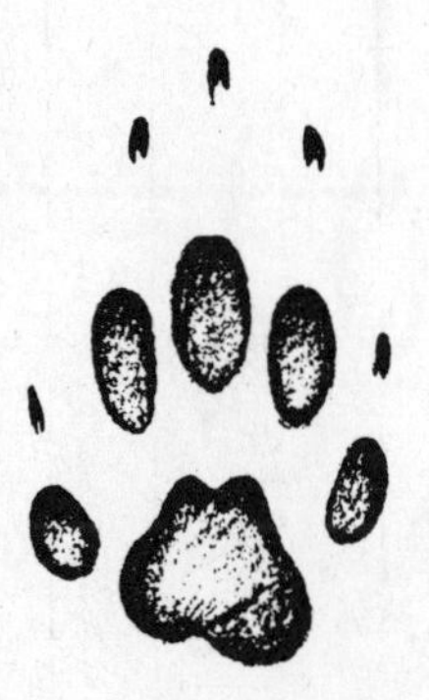

64 Striped Polecat

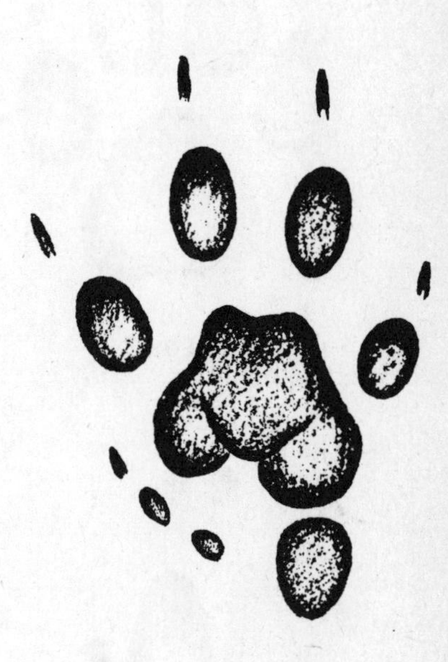

68.1 Large Grey Mongoose

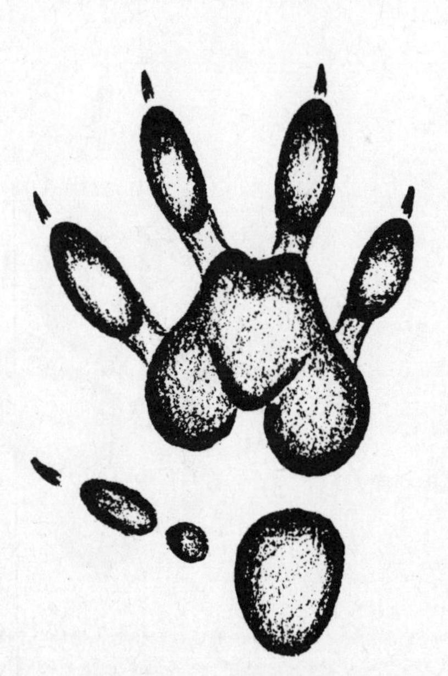

68.4 Water Mongoose

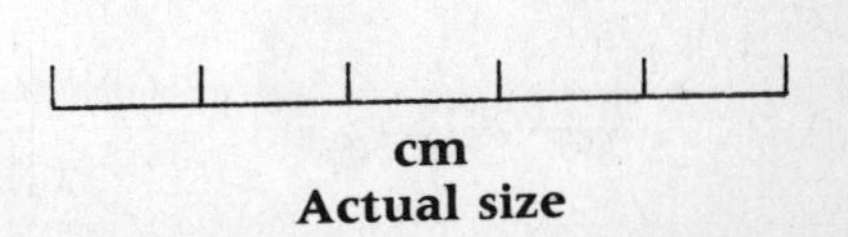

cm
Actual size

69.3 Domestic Dog (Pomeranian)

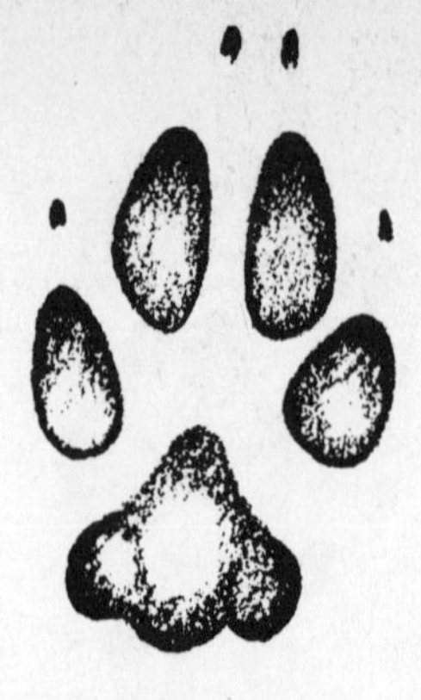

69.2 Cape Fox

70.2 Side-striped Jackal

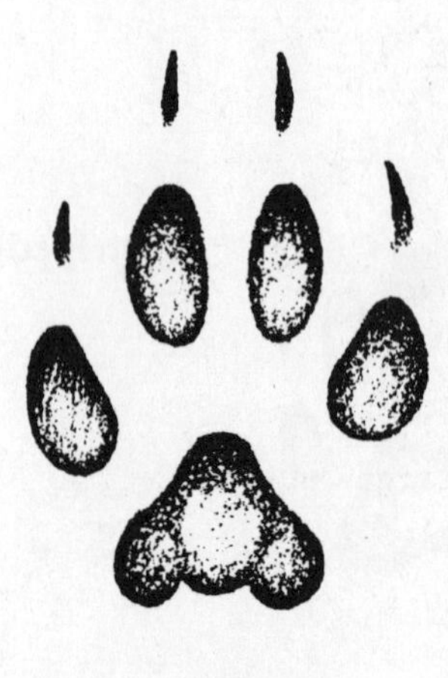

69.1 Bat-eared Fox

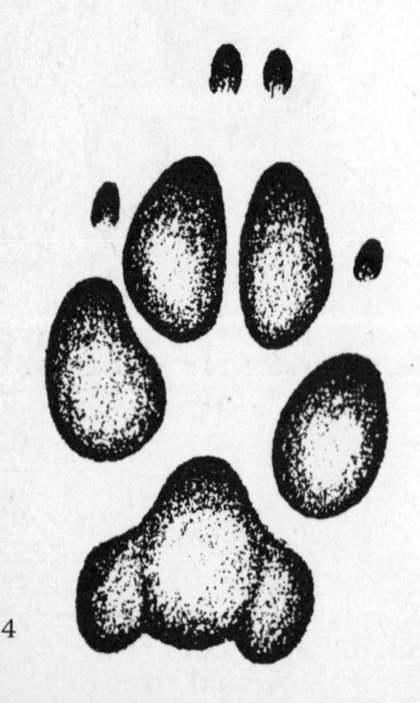

70.1 Black-backed Jackal

72 Aardwolf

cm
Actual size

74.2 African Wild Cat

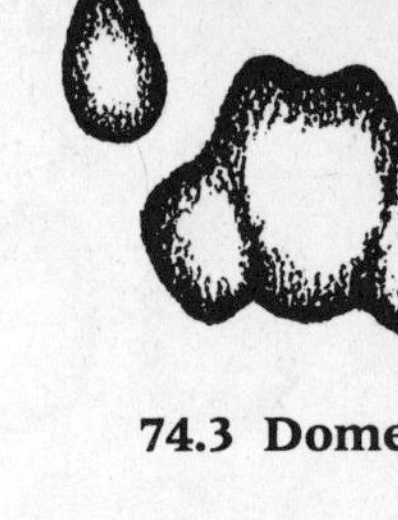

74.3 Domestic Cat

74.1 Small Spotted Cat

67.1 Small-spotted Genet

67.2 Large-spotted Genet

66 Tree Civet

74.5 Caracal

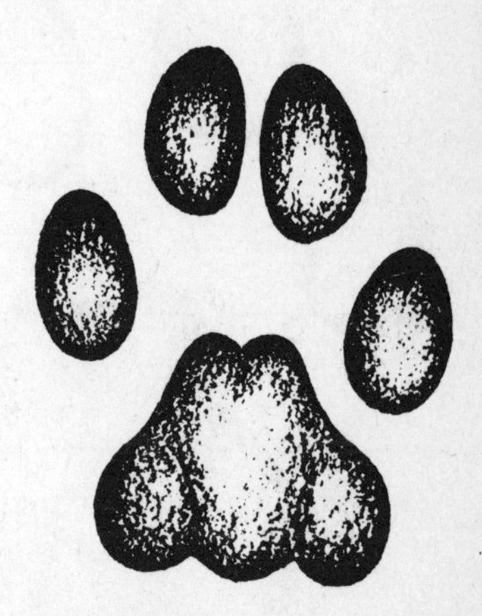

74.4 Serval

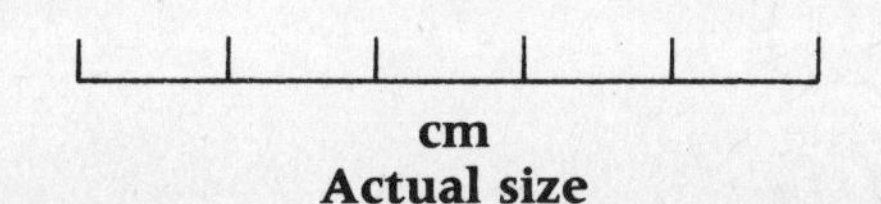

cm
Actual size

65 African Civet

71.1 Wild Dog

75.1 Cheetah

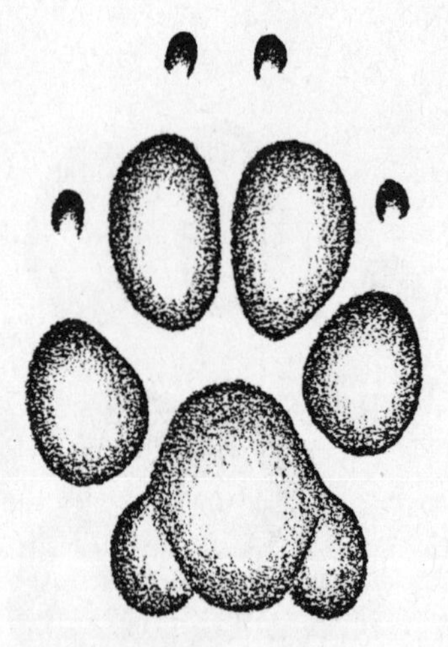

70.3 Domestic Dog (Kalahari tracker dog)

cm
Reduced

Gemsbok and Springbok at waterhole

Buffalo

73.1 Brown Hyaena

73.2 Spotted Hyaena

75.2a Leopard

75.3 Lion

cm
Reduced

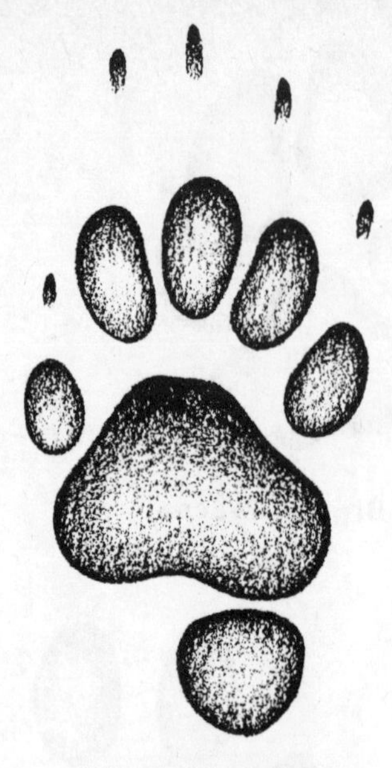

62 Honey Badger

55 Porcupine

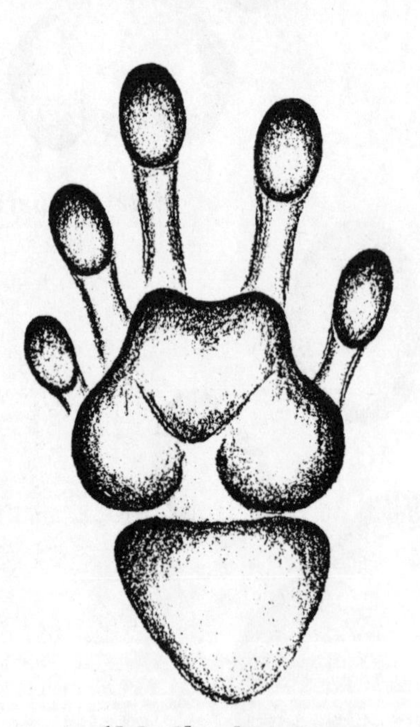

61.1 Clawless Otter

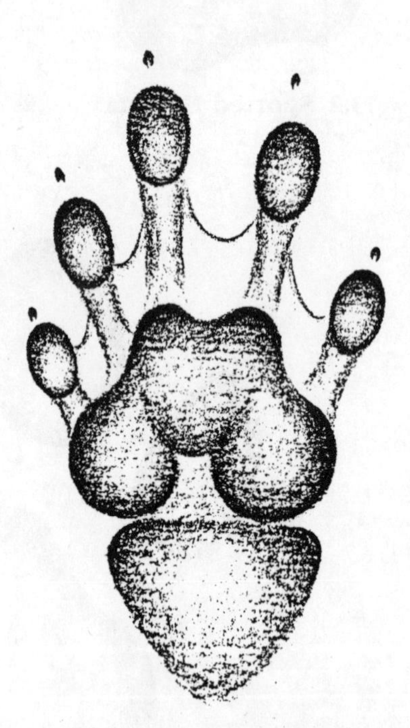

61.2 Spotted-necked Otter

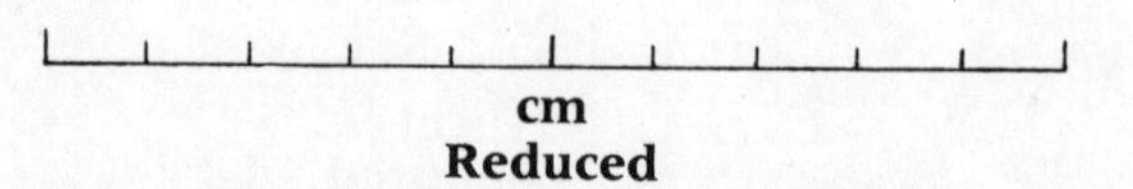

Reduced

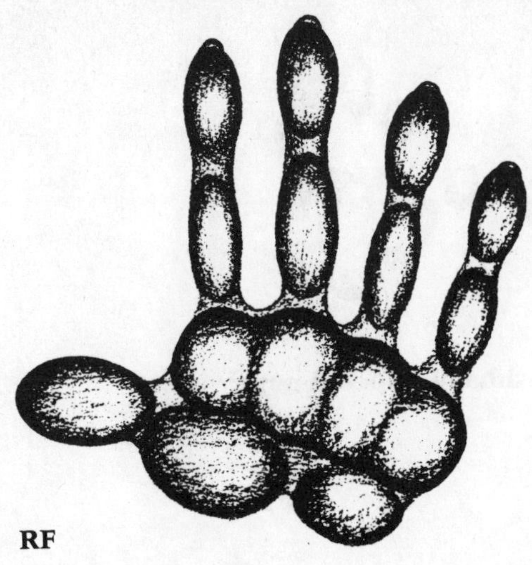

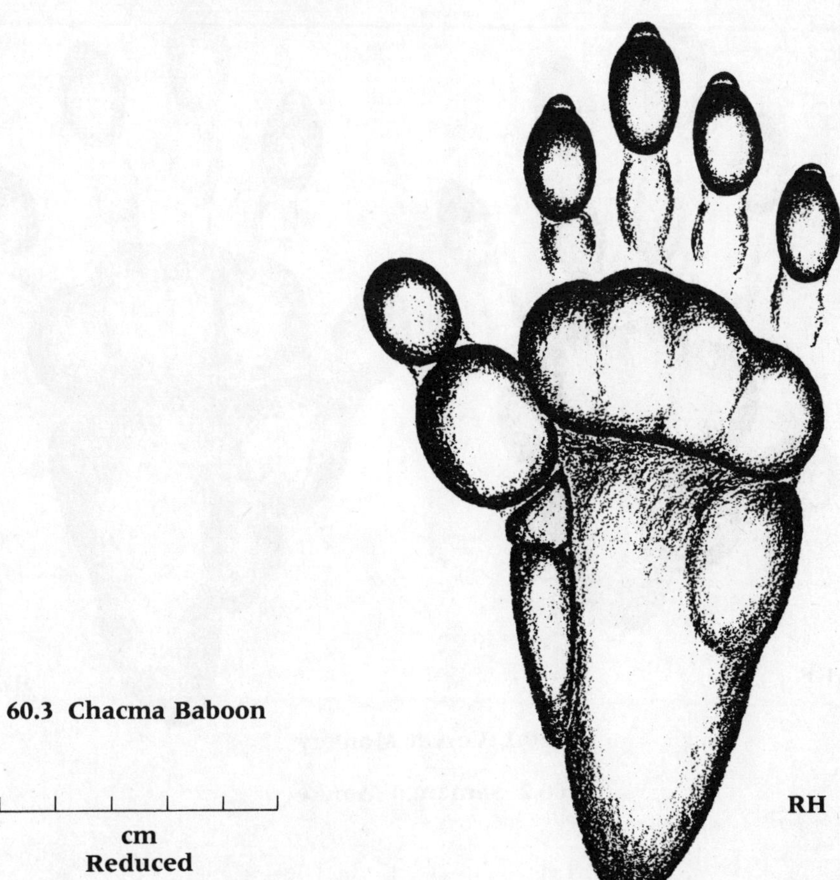

60.3 Chacma Baboon

59.1 Lesser Bushbaby (bipedal hop)

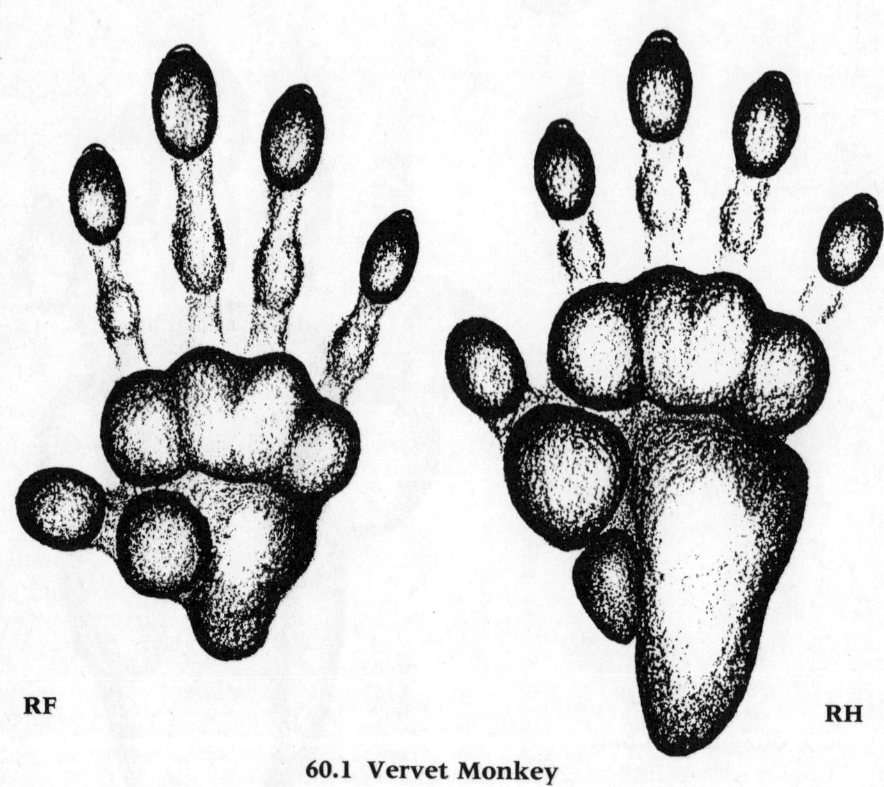

60.1 Vervet Monkey

60.2 Samango Monkey

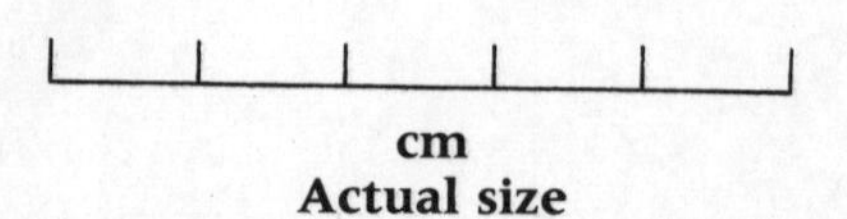

cm
Actual size

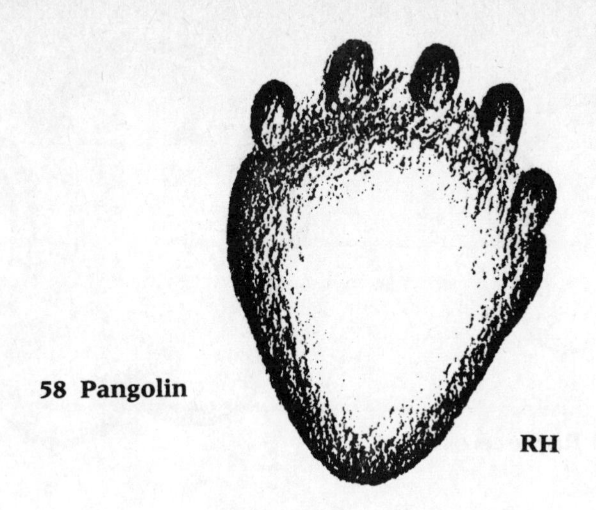

58 Pangolin

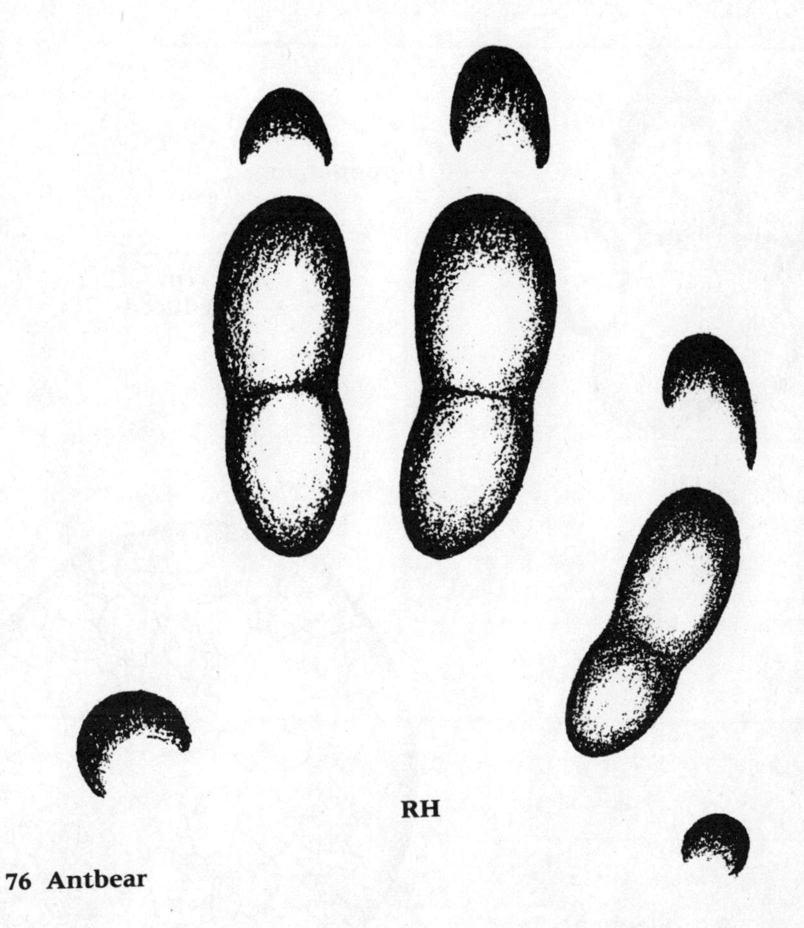

76 Antbear

cm
Actual size

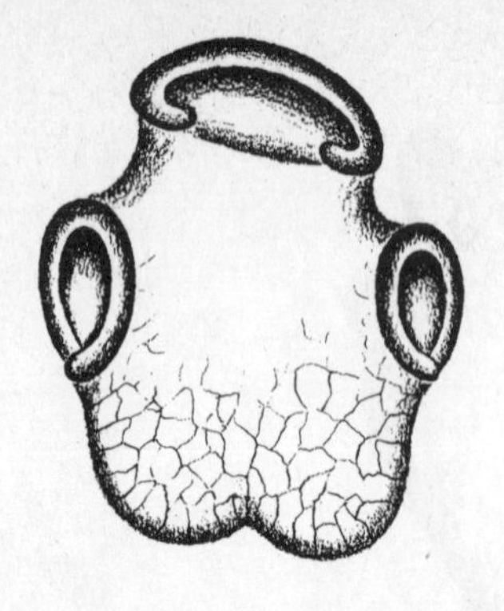

79.2 Hook-lipped Rhinoceros

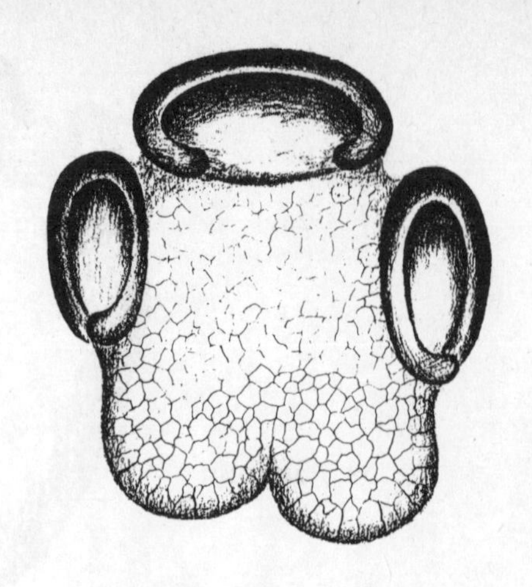

79.1 Square-lipped Rhinoceros

80 Hippopotamus

cm
Reduced

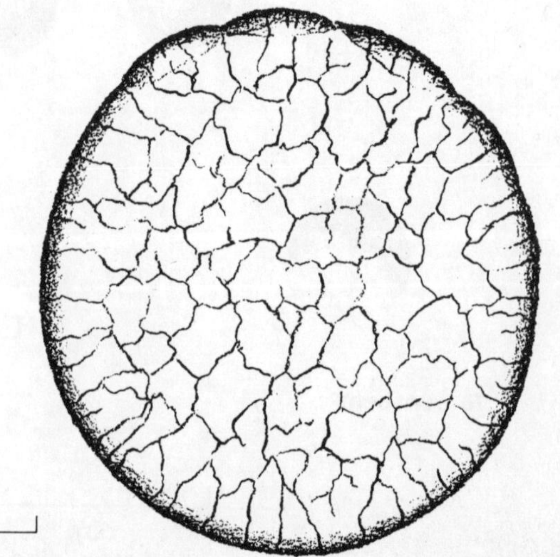

77 Elephant

50 cm

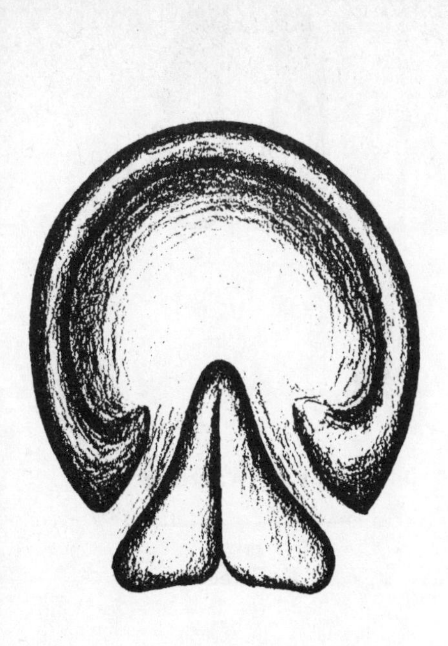

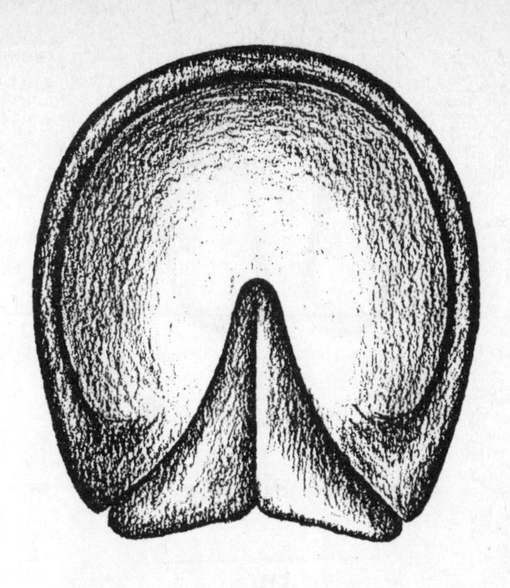

78.1 Burchell's Zebra

78.2 Cape Mountain Zebra

78.3 Hartmann's Mountain Zebra

78.4 Domestic Donkey

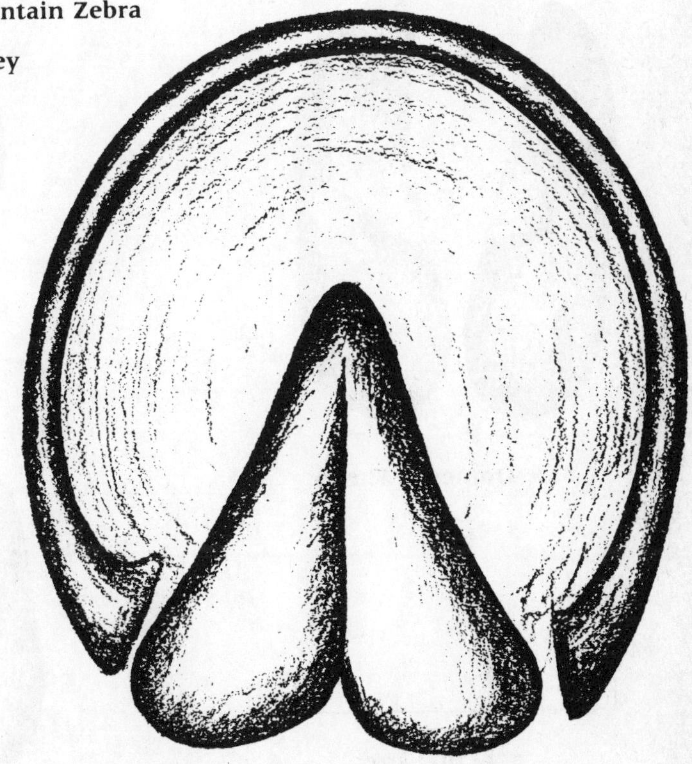

cm
Reduced

78.5 Horse

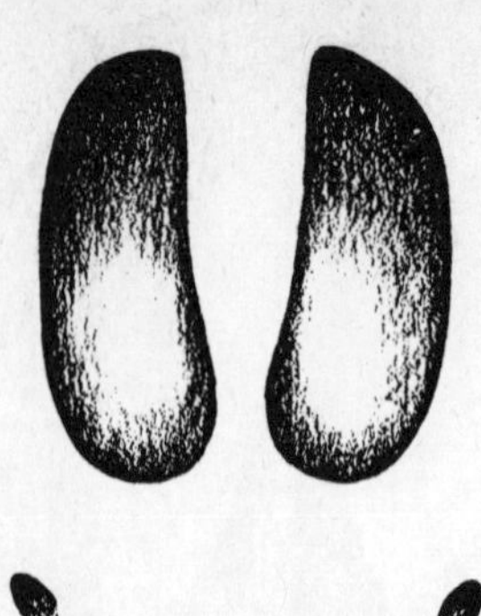

81.2 Bushpig

cm
Reduced

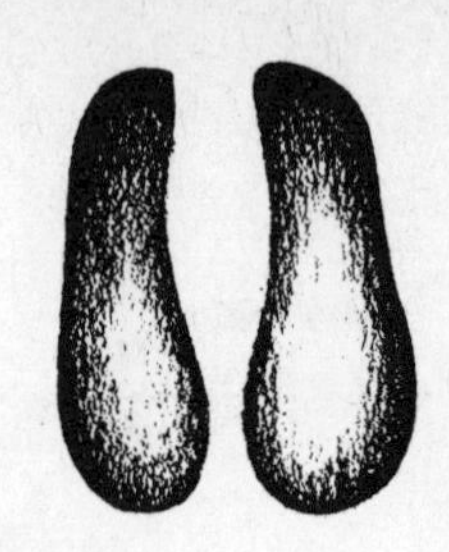

81.1 Warthog

cm
Reduced

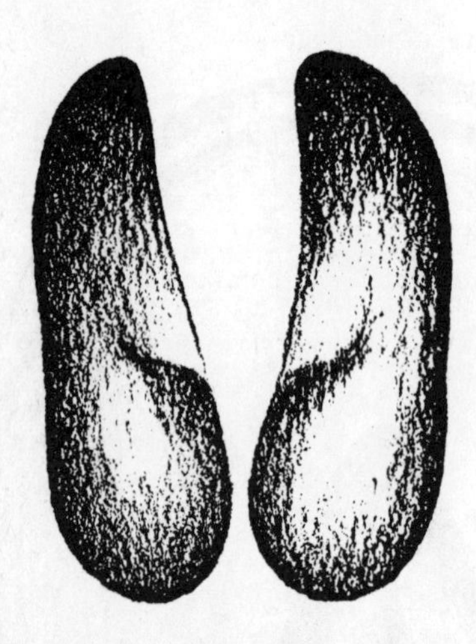

81.5 Domestic Sheep

83 Klipspringer

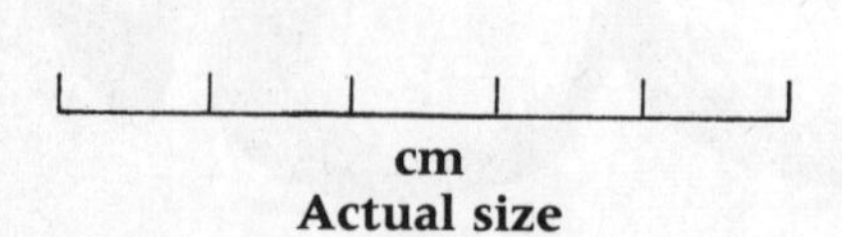

cm
Actual size

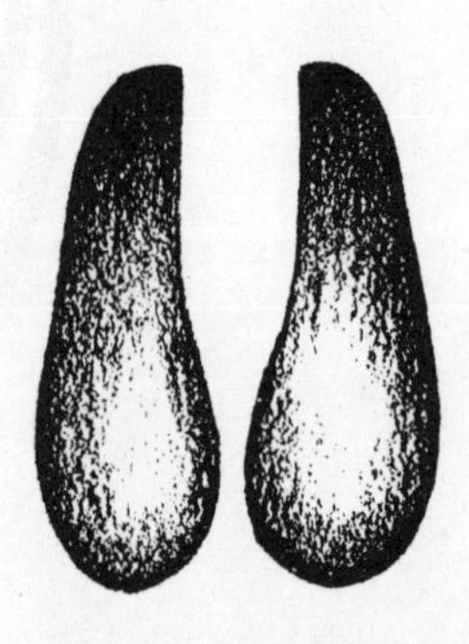

81.4 Domestic Goat

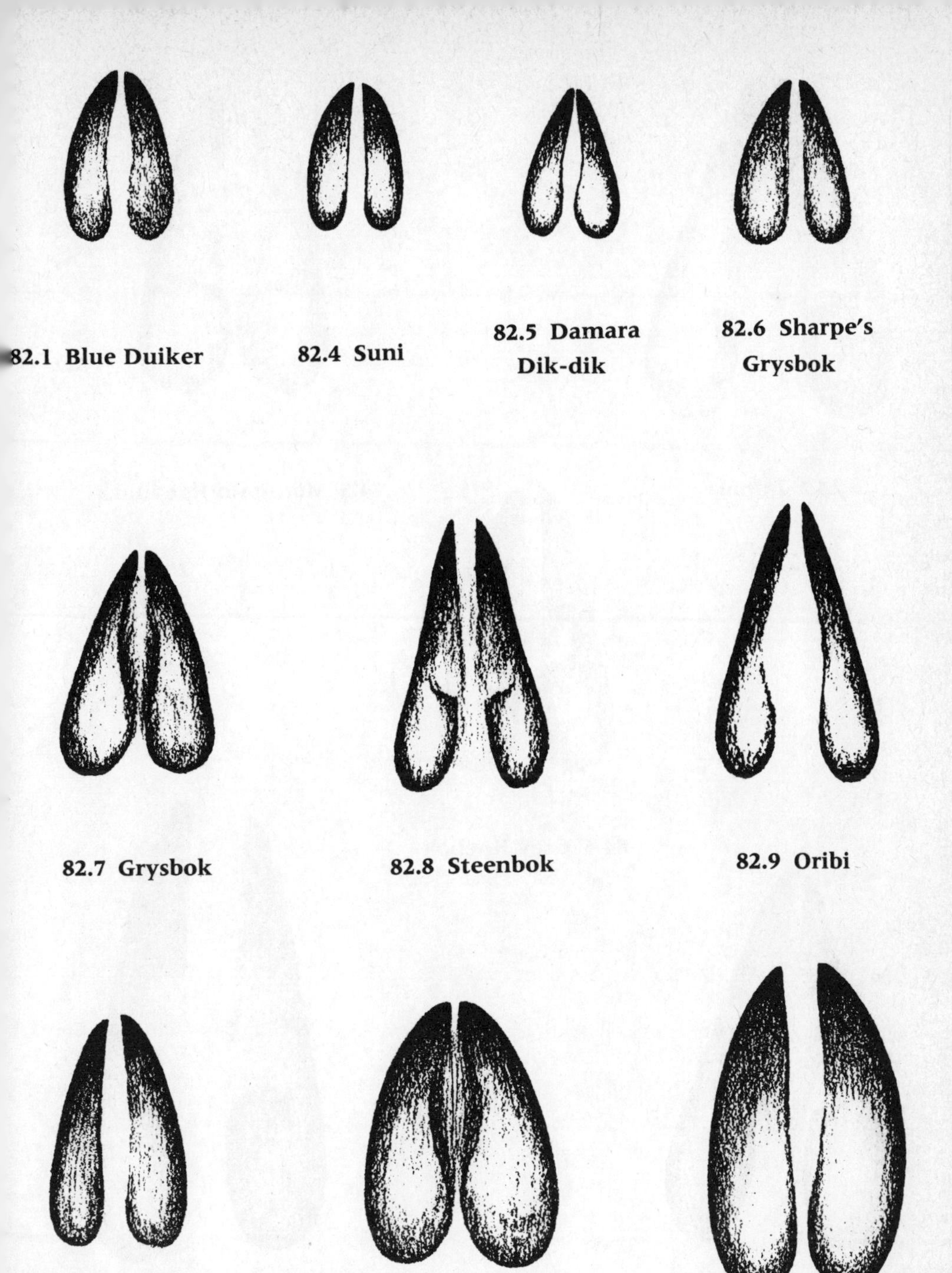

82.2 Red Duiker **82.3 Common Duiker** **87.1 Bushbuck**

cm
Actual size

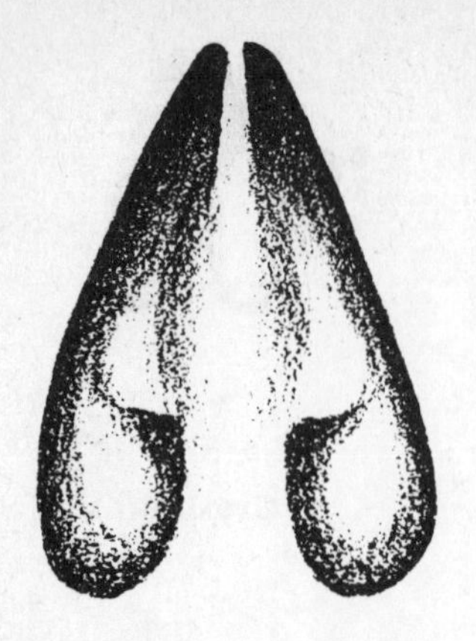

84.2 Impala

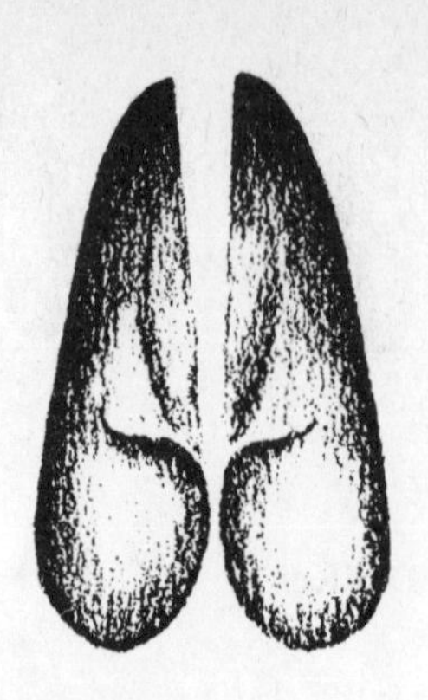

84.5 Mountain Reedbuck

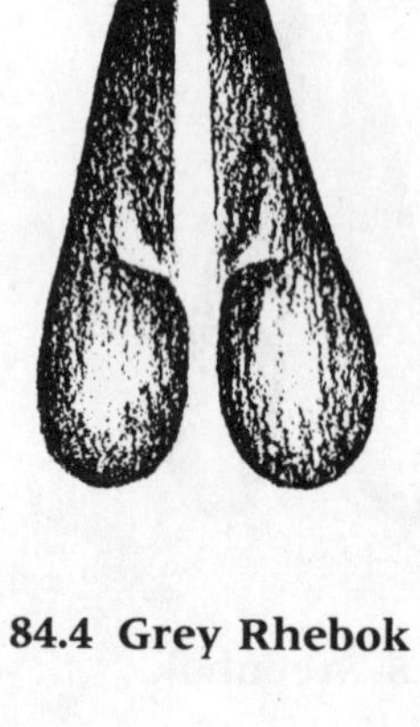

84.4 Grey Rhebok

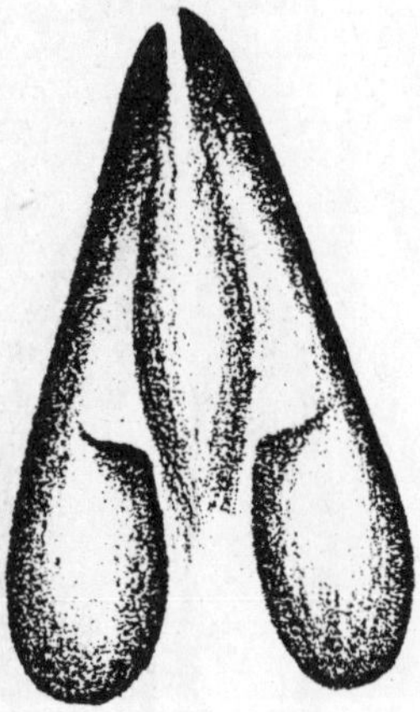

84.1 Springbok

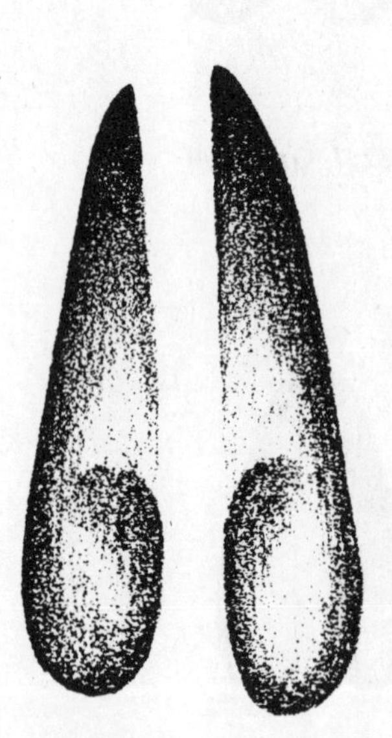

84.6 Reedbuck

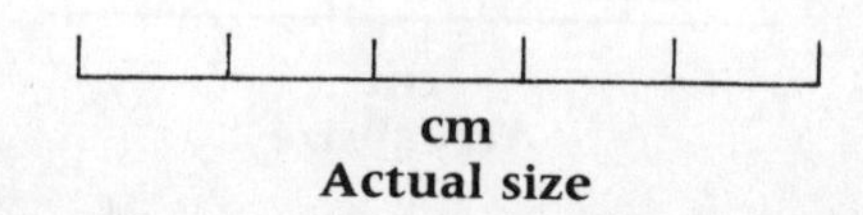

cm
Actual size

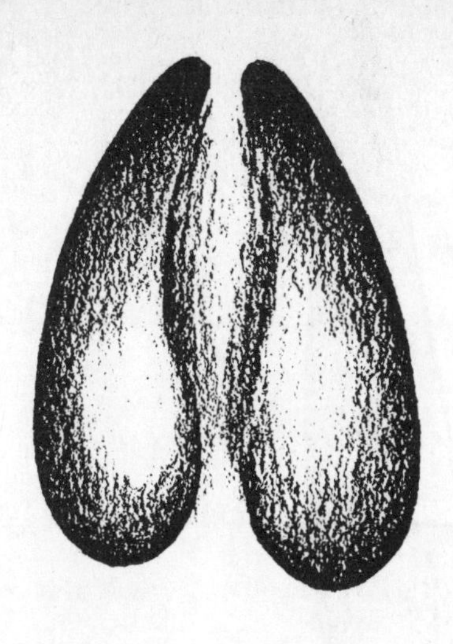

92 European Fallow Deer

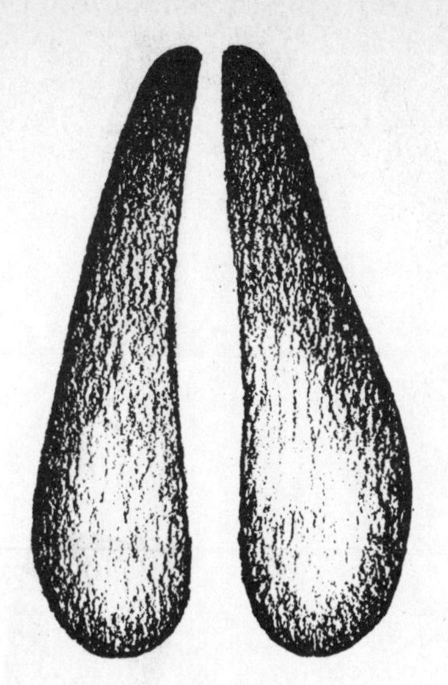

85.2 Puku

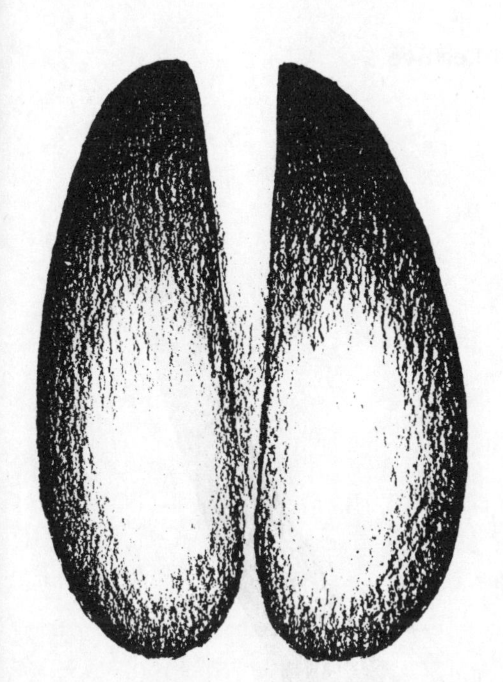

87.3 Kudu

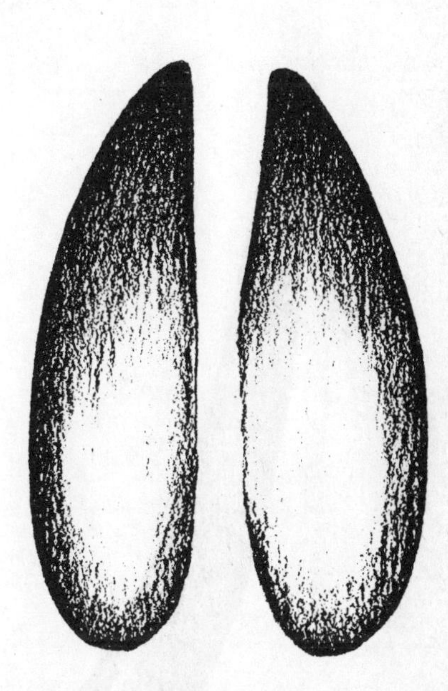

87.2a Nyala

cm
Actual size

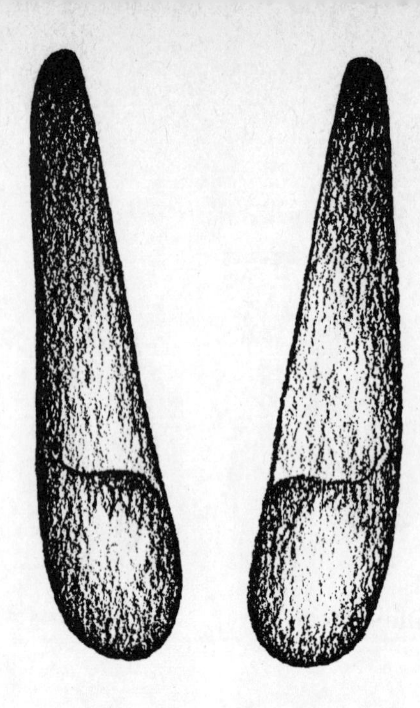

85.1 Red Lechwe

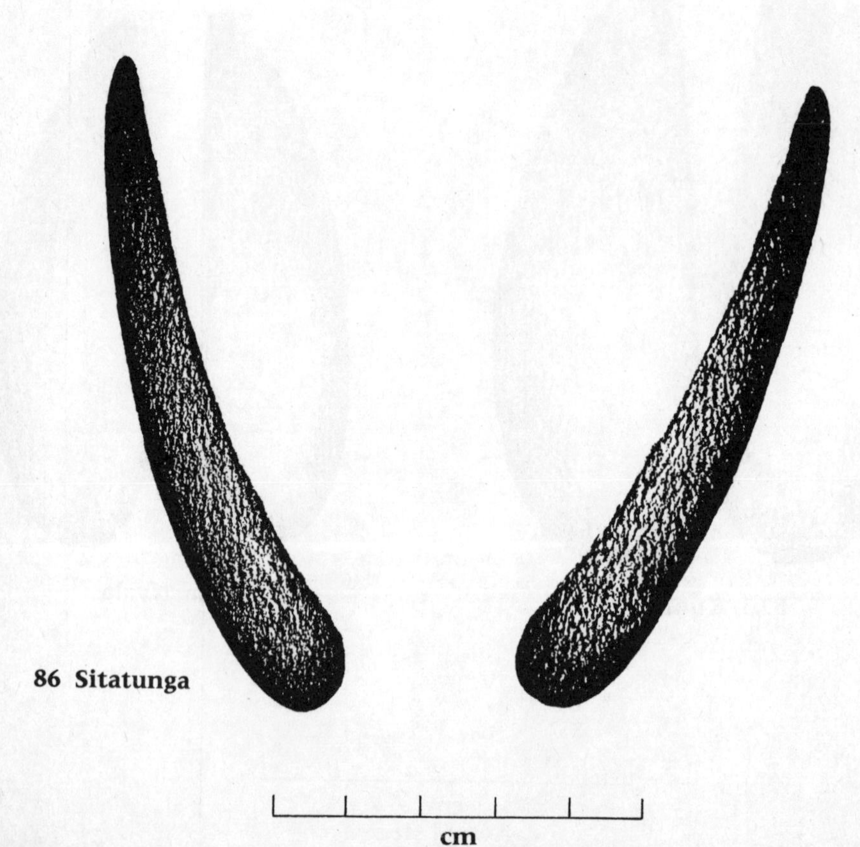

86 Sitatunga

cm
Actual size

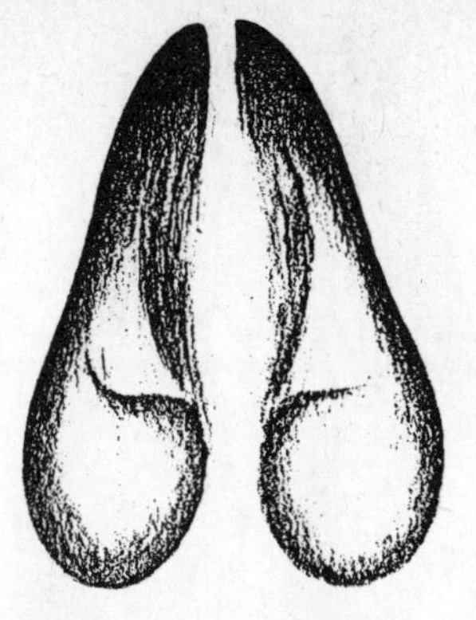
89.4 Bontebok

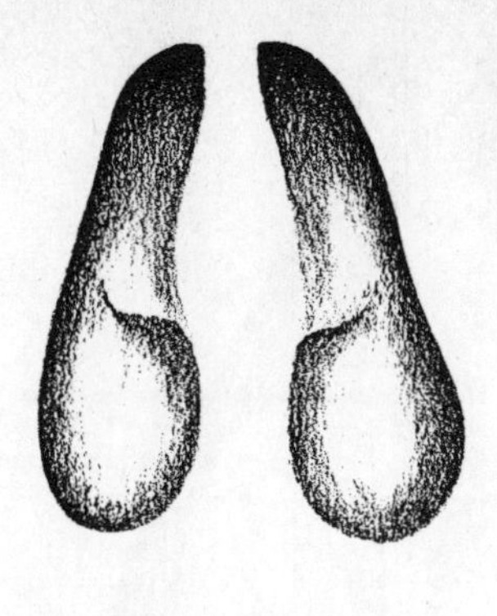
89.5 Blesbok

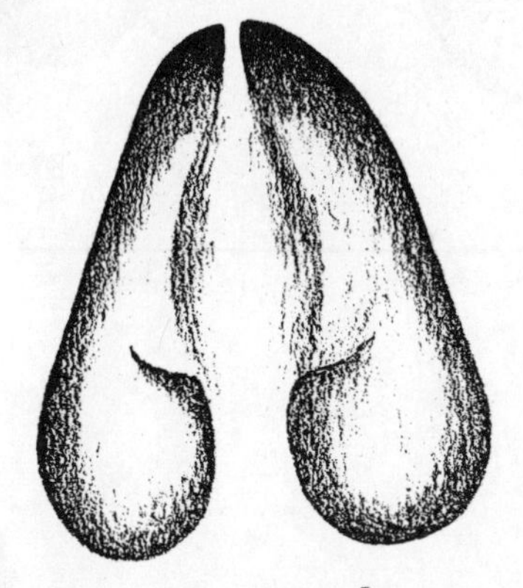
89.6 Tsessebe

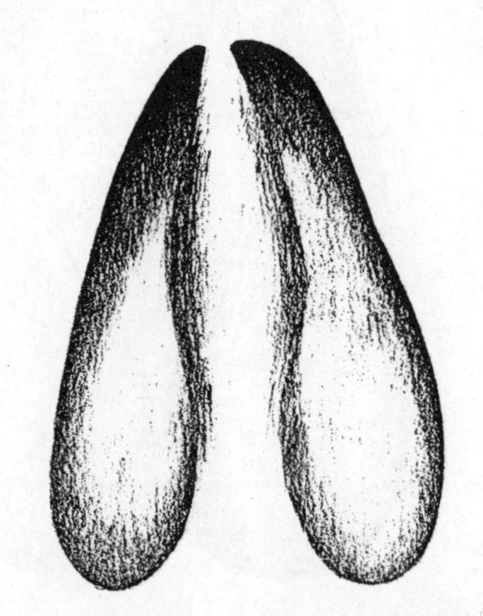
89.8 Lichtenstein's Hartebeest

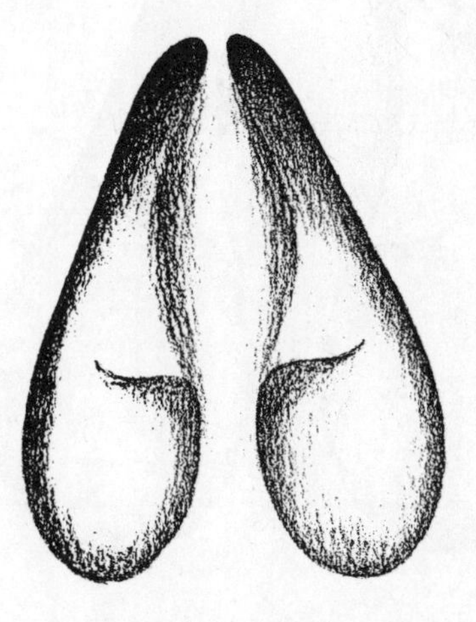
89.7 Red Hartebeest

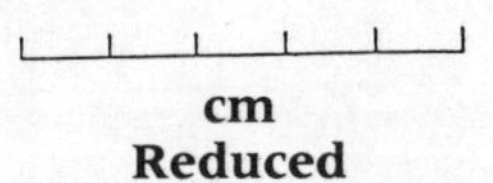
cm
Reduced

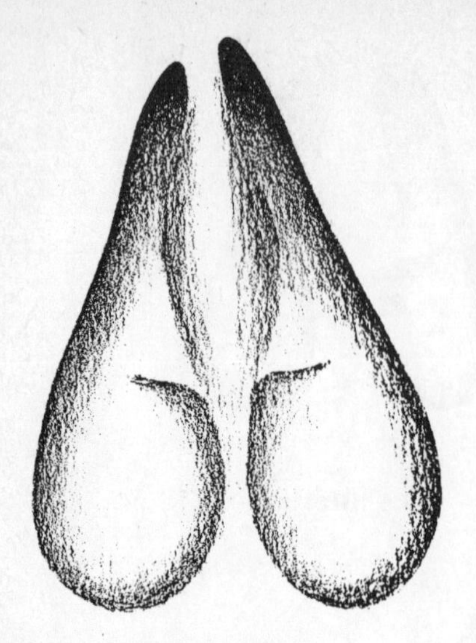

89.2 Sable

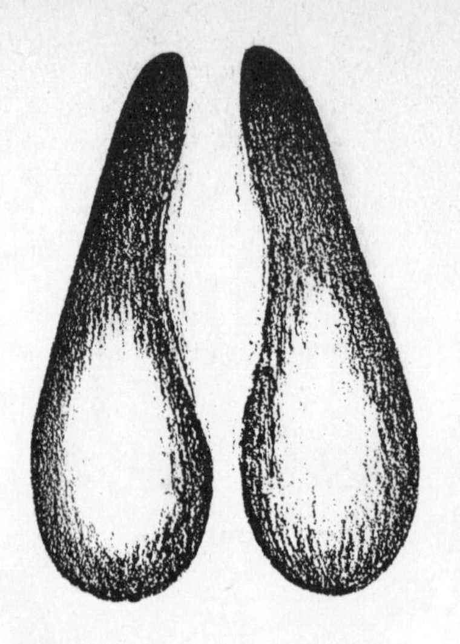

85.3 Waterbuck

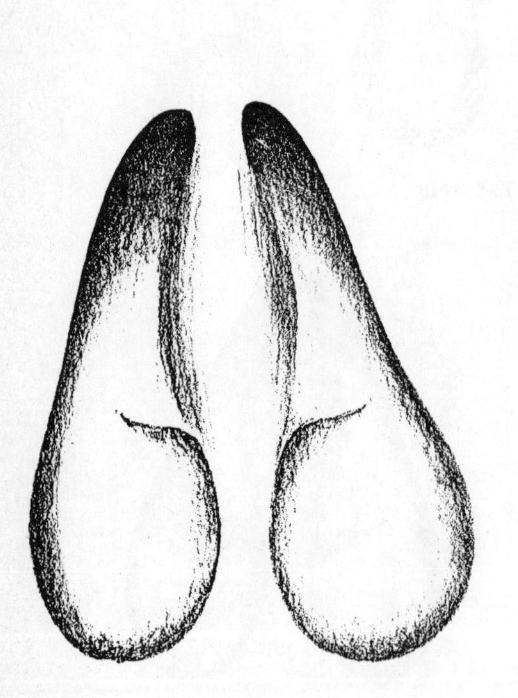

89.3 Roan

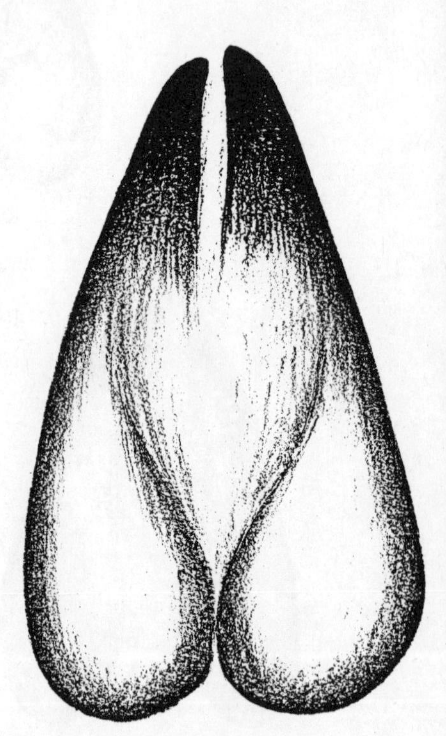

89.1 Gemsbok

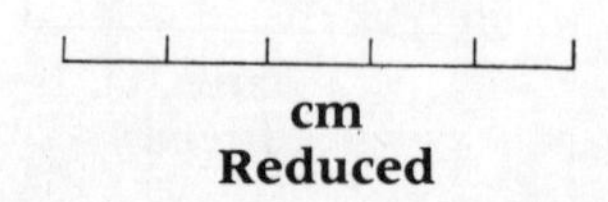

cm
Reduced

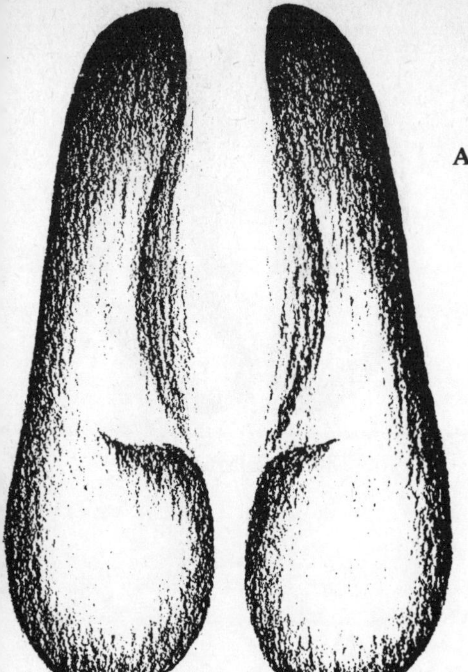

Actual size cm

89.9 Black Wildebeest

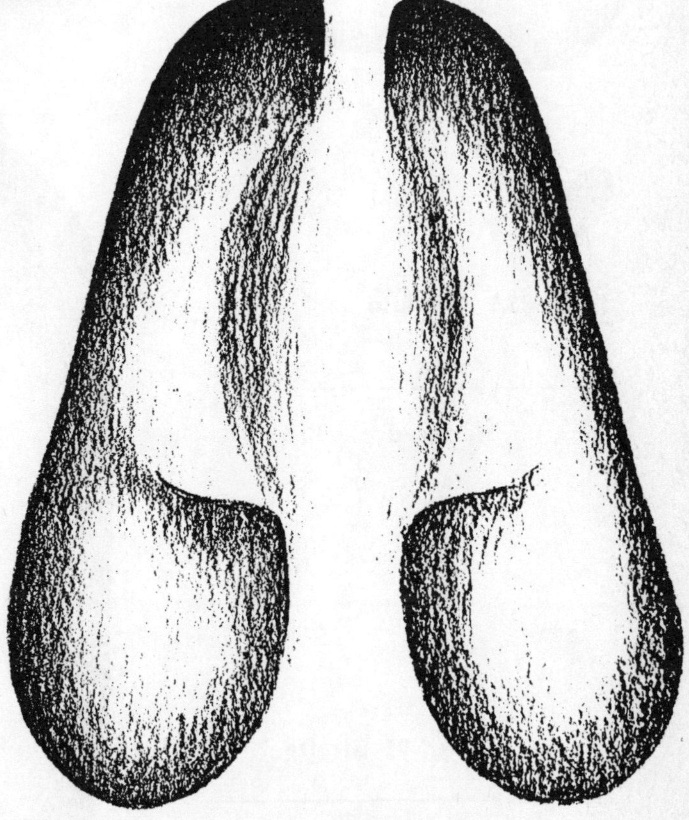

89.10 Blue Wildebeest

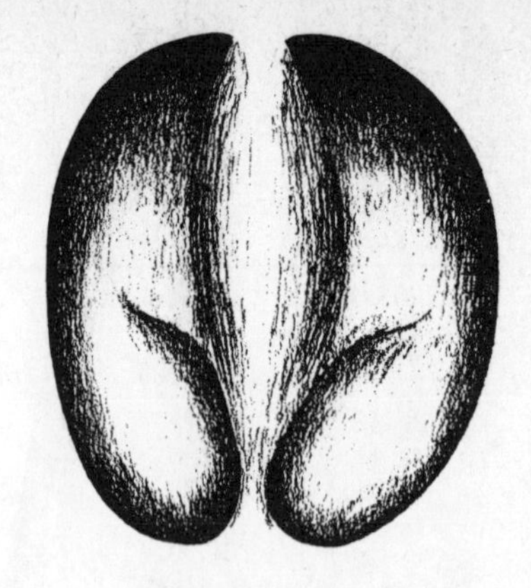

88.1 Eland

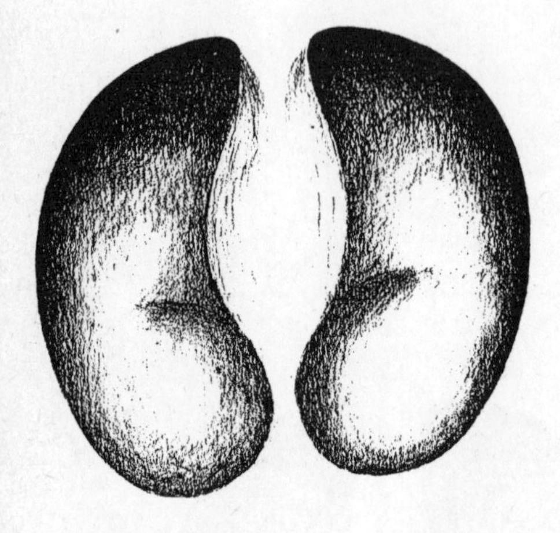

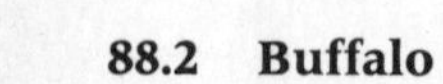

88.2 Buffalo

cm
Reduced

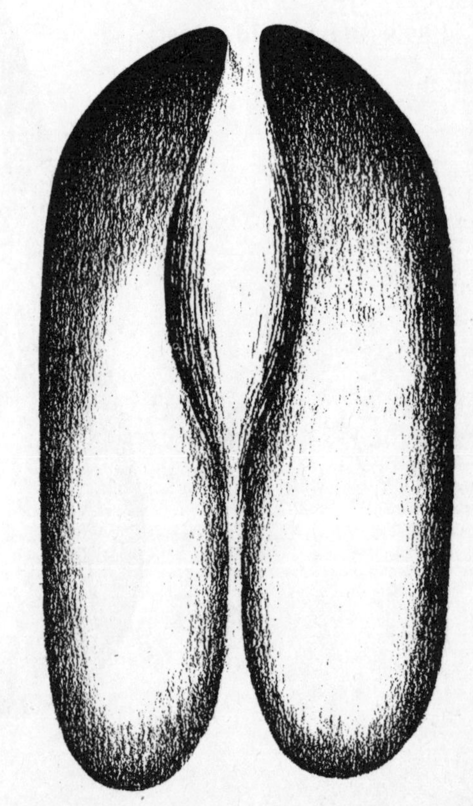

91 Giraffe

cm
Reduced

Kudu

Lion and lioness

Elephant

Burchell's Zebra

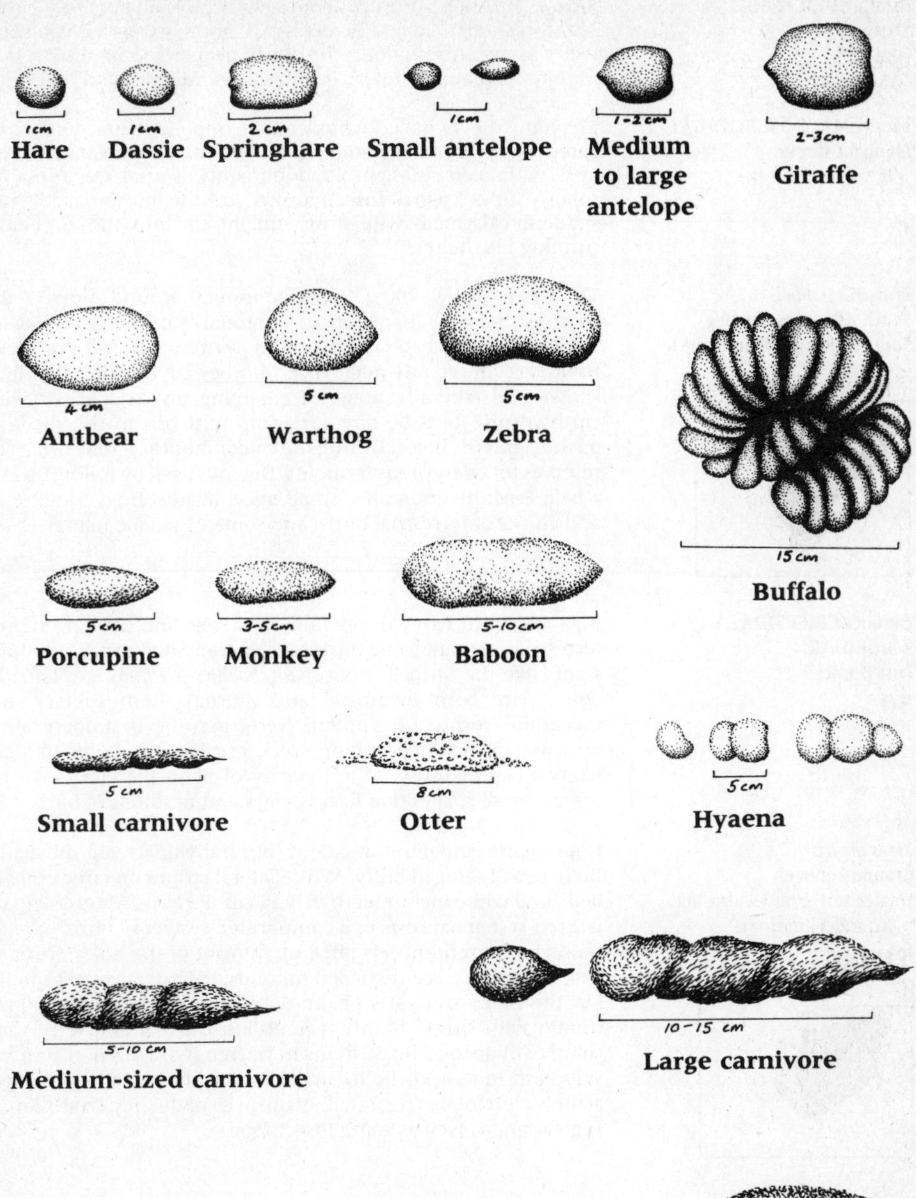
1cm
Hare
1cm
Dassie
2 cm
Springhare
1cm
Small antelope
1-2cm
Medium to large antelope
2-3cm
Giraffe
4 cm
Antbear
5 cm
Warthog
5 cm
Zebra
15 cm
Buffalo
5 cm
Porcupine
3-5 cm
Monkey
5-10 cm
Baboon
5 cm
Small carnivore
8 cm
Otter
5 cm
Hyaena
5-10 cm
Medium-sized carnivore
10-15 cm
Large carnivore

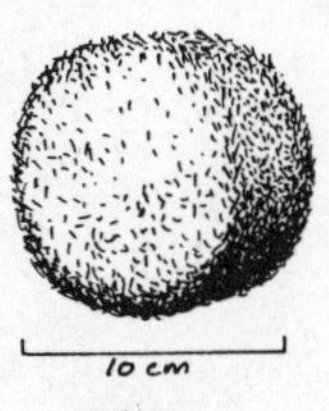
10 cm
Hippo

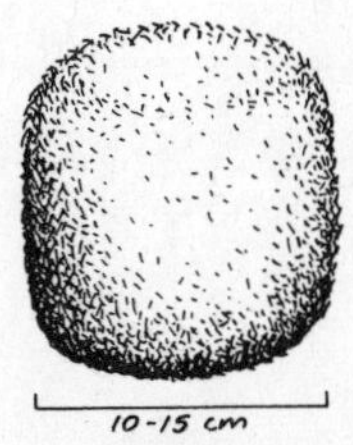
10-15 cm
Rhino

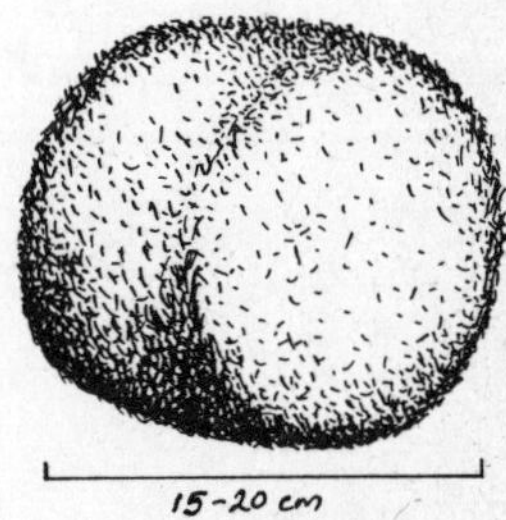
15-20 cm
Elephant

Family SORICIDAE
Shrews
44

Shrews have long, narrow, pointed muzzles and very small eyes. Mainly solitary. Generally terrestrial, but most species can climb well. Active throughout the 24-hour period, in bursts, and require food and water on regular basis. Mainly insectivorous.

Family MACROSCELIDIDAE
Elephant shrews
45

Elephant shews have a characteristic mobile, trunk-like, elongated snout, broad, upstanding, mobile ears, and relatively large eyes, with keen eye-sight. Predominantly diurnal. Occur mainly solitary or in pairs. Insectivorous for the most part. When suddenly alarmed, will jump straight up into the air before running for shelter.

Erinaceus frontalis
South African Hedgehog
Suid-Afrikaanse Krimpvarkie
47

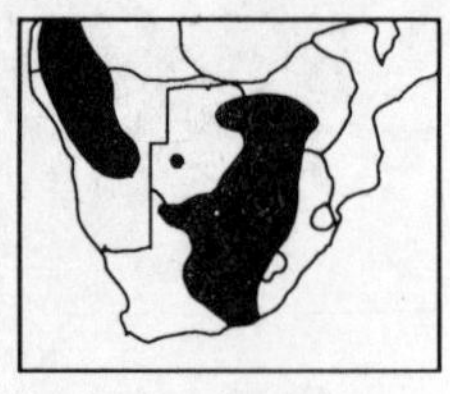

TL 20 cm. Mass 400 g. Usually moves around slowly, the hind-feet leaving characteristic drag marks in the sand. **Habitat:** Wide variety of habitats, including scrub bush and grassland. **Habits:** Predominantly nocturnal. May be active by day after light showers to take advantage of emerging insects and surfacing earthworms. Rests by day curled up in debris in the shade of bushes, grass or holes. During the colder months it relies on its fat reserves for energy requirements. Defends itself by rolling up into a ball. **Food:** Invertebrates, small mice, lizards, frogs, slugs, eggs and chicks of terrestrial birds, and some vegetable matter.

Families CRICETIDAE & MURIDAE
Rats & mice
50

Rats and mice vary considerably in habits and habitat. Mainly terrestrial, though some burrow. Generally nocturnal, although some, like the Striped Mouse (*Rhabdomys pumilio*), are diurnal. Others are both nocturnal and diurnal. Both solitary and gregarious forms are known. Nests may be in underground burrows, piles of vegetation, rock crevices, or holes in tree-trunks. Food consists of a diversity of plant material, invertebrates, small snakes and lizards, eggs and nestlings of birds.

Xerus inauris
Ground Squirrel
Waaierstert-grondeekhoring (Grondeekhoring)
51.1

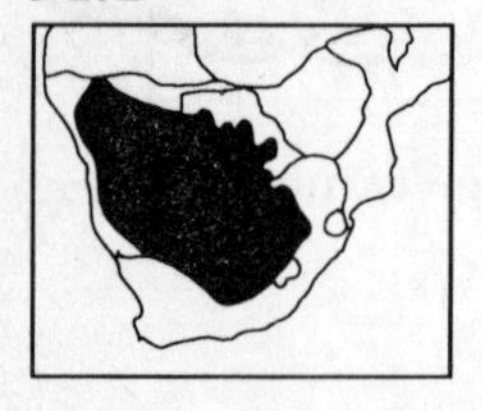

Upper parts cinnamon in colour, but individuals vary in shade. Belly usually tinged buffy. White lateral stripes on either side of body and white incisor teeth. HB 24 cm. T 21 cm. Mass 600 g. Its warren system consists of a complicated system of burrows. Soil removed from funnels is piled up in front of the holes, forming characteristic crescent-shaped mounds. **Habitat:** Occurs throughout the more arid parts of the Subregion. Preference for open terrain with sparse bush cover. **Habits:** Diurnal and terrestrial, colonies of up to about 30 living in warrens with many entrances. Will bask in sun, or lie in shade in very hot weather. Always acutely alert for danger, it will sit up on its haunches. **Food:** Mainly vegetarian, as well as some insect food.

Xerus princeps
Mountain Ground Squirrel
Berg-waaierstert-grondeekhoring (Bergeekhoring)

51.2

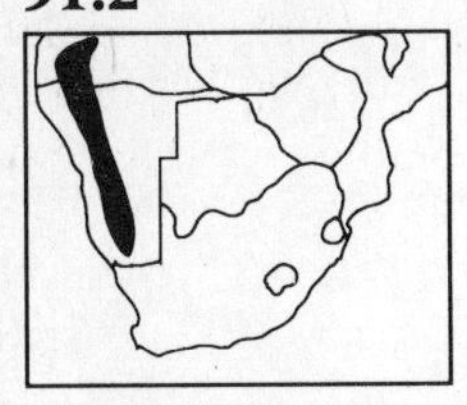

Very similar to Ground Squirrel, *X. inauris*, but slightly lighter in overall colour, and with yellowish incisor teeth. **Habitat:** A rock-dwelling ground squirrel, with warrens frequently on rocky hillsides or among rocky outcrops. The Ground Squirrel, *X. inauris*, which prefers open flat country, normally avoids these rocky areas. **Habits:** Strictly diurnal and lives in warrens. Habits similar to those of the Ground Squirrel. **Food:** Probably same as Ground Squirrel.

Paraxerus cepapi
Tree Squirrel (Bush Squirrel; Yellow-footed Squirrel)
Boomeekhoring

51.3

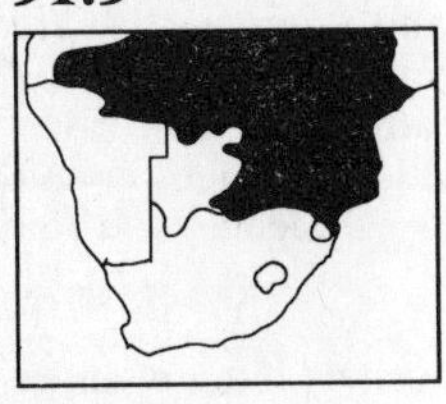

Great variety of colour and size. Has an overall pale grey colour in the western parts of their range, while in the eastern parts is darker, more buffy, sometimes rusty in colour. Under parts vary from white to yellowish or buffy. HB 18 cm. T 17 cm. Mass 200 g for males, females slightly lighter. **Habitat:** Wide variety of types of savanna woodland. **Habits:** Generally solitary, but also occurs in pairs or female with two or three young. Arboreal and terrestrial; much time is spent foraging on the ground. Diurnal, it is most active in the morning and late afternoon. Will bask in the sun when emerging from the nest. Nests are natural holes in trees, lined with leaves or grass. Its alarm call is a high-pitched whistle. **Food:** Predominantly vegetarian, also insects.

Heliosciurus mutabilis
Sun Squirrel
Soneekhoring

51.4

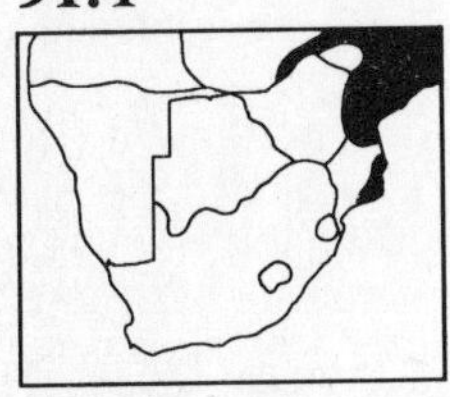

The largest of the arboreal squirrels found in the Subregion. HB 23 cm. T 27 cm. Mass 400 g. Upper parts are grizzled light brown in colour, but colour varies. The tail has narrow light and broad dark bands. **Habitat:** Lowland or montane evergreen forest, but also riverine forest and thickets within woodland. **Habits:** Solitary or in pairs. Most active in late morning and afternoon. Rests in holes in trees or secluded, sheltered places such as dense clumps of creepers high in forest trees. **Food:** Mainly vegetarian, also some insects.

Funisciurus congicus
Striped Tree Squirrel
Gestreepte Boomeekhoring

51.5

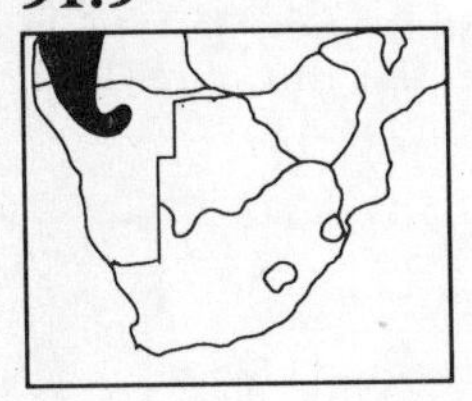

The smallest squirrel in the Subregion. HB 15 cm. T 17 cm. Mass 110 g. Has distinct lateral white stripes along the mid-body, with a dark band below these stripes. The upper parts above the white stripes are dark buffy-yellow, the under parts whitish. **Habitat:** Like the Tree Squirrel (*P. cepapi*), it is associated with woodland, but more closely confined to the denser types, where there are larger trees with more luxuriant canopies. **Habits:** Arboreal, but spends almost as much time foraging on the ground. Lives in small family parties. Diurnal, being most active in early morning. Rests in holes in trees that are lined with leaves and grass or in dreys constructed of twigs, leaves and grass in the forks of branches. The alarm call is a bird-like chirping, while flicking the tail, or a high-pitched whistle-like chattering. **Food:** Predominantly vegetarian, it also eats insects.

Paraxerus palliatus
Red Squirrel
Rooi Eekhoring
51.6

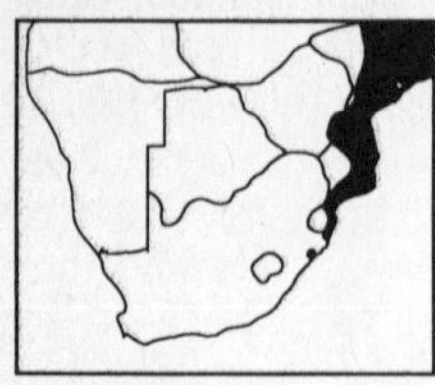

Upper parts and tail are reddish or auburn, although colour varies. HB 20 cm. T 20 cm. Mass 300 g. **Habitat:** Dry or moist evergreen forests, woodlands or riverine or other thickets, where these have a shady understorey of leafy vegetation. **Habits:** Predominantly arboreal, but forages on the forest floor. Diurnal. Generally solitary, or female with young, or male with female while she is in oestrus. **Food:** Mainly nuts, berries and wild fruits.

Sciurus carolinensis
Grey Squirrel
Gryseekhoring
51.7

Introduced to parts of the southwestern Cape. Summer coat is smoother and yellowish-brown in colour, and tail hair sparser than in winter coat. Pelage of winter coat is dense and silvery-grey, tail dark grey with white fringe, and under parts pure white. HB 28 cm. T 22 cm. Mass 580 g. **Habitat:** Sufficient numbers of its staple food-tree, the oak, *Quercus robur*, and three species of pines, *Pinus pinea*, *P. pinaster* and *P. canariensis*, are essential. **Habits:** Solitary or female with young. Arboreal. Diurnal, most active in early morning and late afternoon. Constructs dreys of twigs, leaves and soft debris, lined with leaves or soft material, or rests up in holes in trees lined with leaves and soft debris. **Food:** Vegetarian.

Petromus typicus
Dassie Rat
Dassierot
52

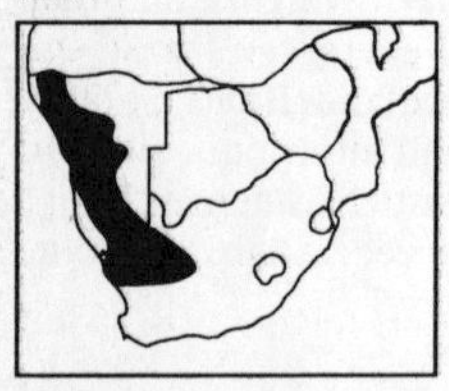

Squirrel-like in appearance, but tail is not bushy. Colour of upper parts from pale grizzled grey to dark chocolate. Under parts paler. HB 17 cm. T 14 cm. Mass 220 g. Spoor probably similar to that of a squirrel. **Habitat:** Closely confined to rocky outcrops, rocky hills and koppies, living in crevices or amongst boulders. **Habits:** Crevices occupied by pairs or family parties. Diurnal, most active early morning or late afternoon. Suns itself when not foraging. **Food:** Vegetarian.

Thryonomys swinderianus
Greater Canerat
Groot Rietrot
53.1

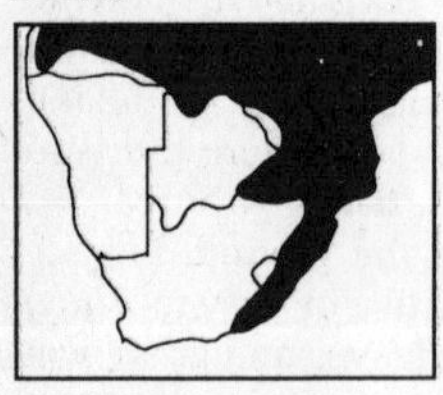

A short-bodied, bulky animal. HB 50 cm. T 20 cm. Mass ♂ 4,5 kg, ♀ 3,6 kg. Body colour speckled dark brown, under parts greyish-white or whitish. **Habitat:** Mainly reedbeds or areas of dense, tall grass with thick reed or cane-like stems. Vicinity of rivers, lakes and swamps, never far from water. **Habits:** Generally solitary, but small groups of up to 10 live in restricted areas of reedbed. Predominantly nocturnal. Distinct runs formed in reedbeds and grass. Resting places in densest part of reedbed. **Food:** Vegetarian.

Pedetes capensis
Springhare
Springhaas
54

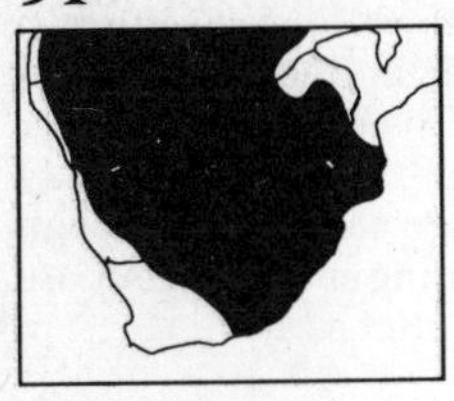

With short front legs, long powerful hind-legs and long tail, it resembles a small kangaroo. The generally cinnamon-buff coloured coat varies in colour from one area to another. Tail has broad black tip, and under parts are whitish. HB 39 cm. T 40 cm. Mass 3,1 kg. It hops on the back feet, holding the front legs close to the body. **Habitat:** Occurs on open sandy ground, sandy scrub, overgrazed grassland, fringes of vleis and dry riverbeds, flood-plain grassland, cultivated areas of open scrub. **Habits:** Nocturnal. Congregates in open groups when feeding. Simple burrows may be extended to include other side burrows and can have several entrances and escape holes. **Food:** A grazer, living mainly on grass.

Hystrix africaeaustralis
Porcupine
Ystervark
55

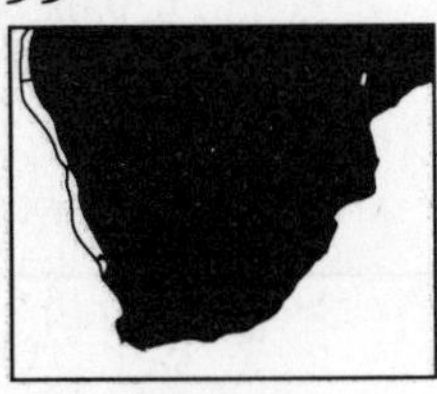

The largest African rodent, characterised by the erectile spines and quills which serve as a means of self-defence. TL 84cm. Mass 18 kg. **Habitat:** Occurs in most types of vegetation, but generally absent from forest. Prefers broken country with rocky hills and outcrops. **Habits:** Usually solitary, occasionally in pairs or female with young. Almost exclusively nocturnal, but will sunbathe close to shelter. It is a noisy animal, snuffling, grunting and rasping its quills and spines against obstacles. Can become aggressive if cornered. To defend itself, runs backwards or sideways so that its sharp quills penetrate the skin of its adversary. **Food:** Predominantly vegetarian, including bulbs, tubers, roots, fallen wild fruits.

Lepus saxatilis
Scrub Hare
Kolhaas
56.1

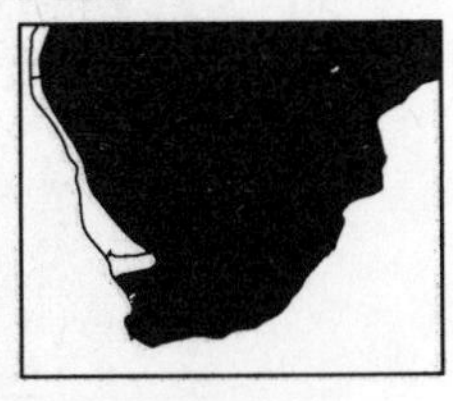

Upper parts are grizzled greyish or buffy. Has a thick mass of hair underneath its feet, so footprints do not show well-defined pad imprints. **Habitat:** Savanna woodland and scrub with a grass cover, but absent from forest, desert and open grassland. **Habits:** Nocturnal. During the day it lies up in forms under bushes where there is some grass cover. Normally solitary, but female in oestrus may be accompanied by one or more males. **Food:** Leaves, stems and rhizomes of dry and green grass.

Lepus capensis
Cape Hare
Vlakhaas
56.2

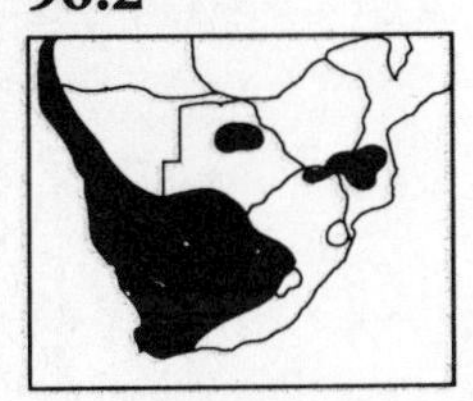

There exists a wide variation in colour and size. In the southwestern Cape it is light buffy in colour, grizzled with black ticking, while in Botswana it is a much lighter whitish-grey with greyish ticking. **Habitat:** Has a preference for a dry, open habitat, such as open grassland plains. As the Cape Hare is a grazer, palatable grasses are essential, as well as cover in the form of clumps of grass in which to lie up. **Habits:** Predominantly nocturnal, but may forage during the day in overcast weather. Lies up during day in forms situated in grass clumps or under small bushes, always alert for danger. Normally solitary, but female in oestrus may be accompanied by several males. Eyesight and hearing acute. **Food:** A grazer, with a preference for areas with short grass, green grass and fresh green shoots.

Genus Pronolagus
Red rock rabbits

Apparently all three species of *Pronolagus* have similar habits and habitat requirements. They are closely confined to areas with substantial shelter in the form of rocks, occurring on krantzes, rocky hillsides, boulder-strewn koppies, in rocky ravines and on piles of rocks in dry river-beds. They lie up during the day in forms under rock ledges or boulders, where their coloration makes them difficult to see. Their habitat must also provide some cover of bushes and palatable grasses. They are all predominantly nocturnal. They are gregarious, but forage solitarily, although numbers may congregate on preferred feeding grounds. They are all grazers, partial to flushing green grass after a burn.

Pronolagus rupestris
Smith's Red Rock Rabbit
Smith se Rooiklipkonyn
56.3

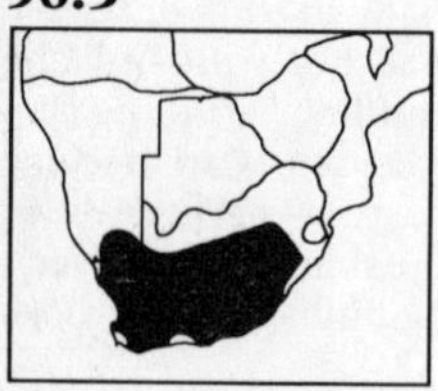

The smallest of the three species of *Pronolagus*, with HB 45 cm and mass 1,6 kg. Colour varies, with upper parts generally rufous-brown with distinct black grizzling. Rump and back of hind-legs are brighter rufous in colour. Tail is bushy with black tip.

Pronolagus crassicaudatus
Natal Red Rock Rabbit
Natalse Rooi Klipkonyn
56.4

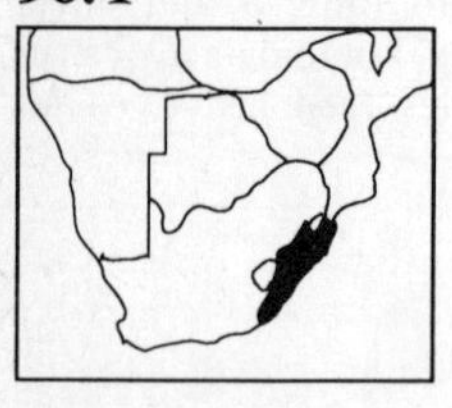

Upper parts grizzled rufous-brown with blackish wash. Rump and back of hind-legs are brighter rufous. Tail is short, not bushy, and uniformly ochraceous-brown. HB 50 cm. Mass 2,6 kg.

Pronolagus randensis
Jameson's Red Rock Rabbit
Jameson se Rooi Klipkonyn
56.5

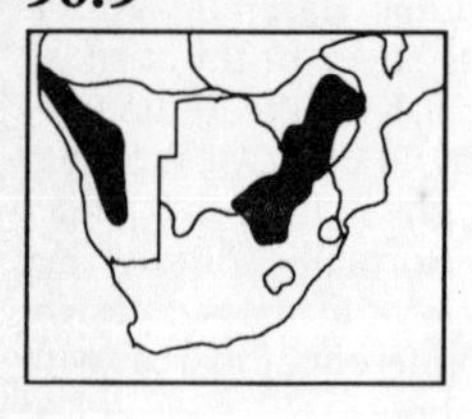

Upper parts grizzled rufous-brown. Rump and back of limbs lighter in colour. Tail large and bushy, and uniform ochraceous-brown with black tip. HB 46 cm. Mass 2,3 kg.

Bunolagus monticularis
Riverine Rabbit
Rivierkonyn
56.6

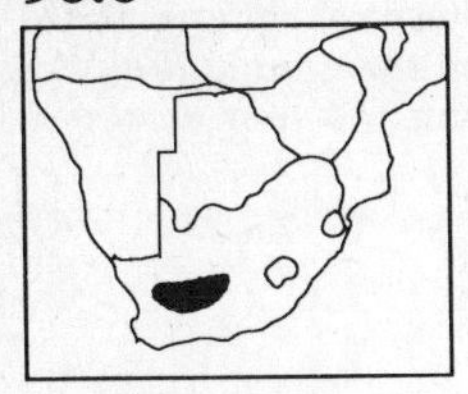

Ranks among the rarest of southern African mammals, and has only been observed a few times since its discovery in 1902. Has distinctive long ears, and characteristic dark brown band along the sides of the lower jaw. Upper parts grizzled drab-grey. Eyes encircled with distinct white rings. Round fluffy tail is pale greyish-brown. HB 43 cm. Mass 1,0–1,5 kg. **Habitat:** Dense riverine bush, as found on the Fish and Rhinoceros rivers in the Karoo. **Habits:** Little is known about their habits.

Procavia capensis
Rock Dassie
Klipdas
57.1

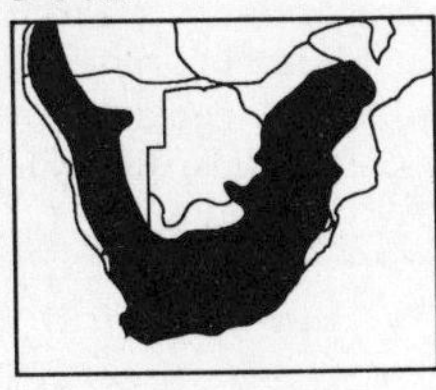

Male has a TL of 50 cm and mass 4,5 kg, females smaller and lighter. Upper parts vary in colour from yellowish-buff to reddish or greyish-brown, with grizzled appearance. **Habitat:** Occurs only in rocky outcrops such as krantzes, rocky koppies or hillsides, or piles of loose boulders with an association of bushes and trees that provide browse. **Habits:** Predominantly diurnal. Gregarious, size of colonies may vary from a family party of 4 to 6 on an isolated pile of rocks to hundreds on extensive ranges of krantzes. When danger threatens, one of the females will give the warning call, when all will dive for shelter. **Food:** Depending on the relative availability of food sources, it is a browser in some parts and predominantly a grazer in other parts.

Manis temminckii
Pangolin
Ietermagog
58

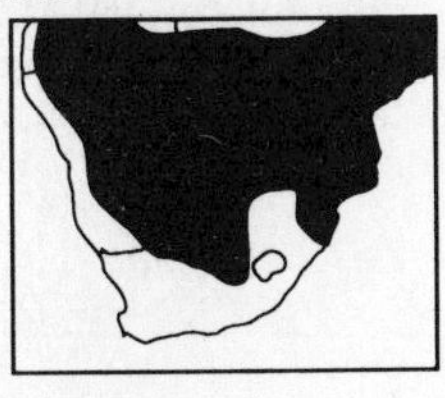

Unmistakable, armoured with scales, reaches an overall length of over 1 metre and mass about 15 kg. **Habitat:** A savanna species, not occurring in forest or in desert. Occurs in scrub and various types of savanna woodland, floodplain grassland, rocky hills and sandveld. Habitat must provide the species of ants and termites it lives on. **Habits:** Solitary. Predominantly nocturnal with some diurnal activity. Under severe stress it will curl up into a tight ball, the tough scales protecting the head and soft under parts. By day it shelters in holes, such as disused Antbear or Springhare burrows, or hides in piles of debris in the shade. **Food:** Ants and termites, including larvae and pupal stages.

Galago moholi
Lesser Bushbaby (Night Ape)
Nagapie
59.1

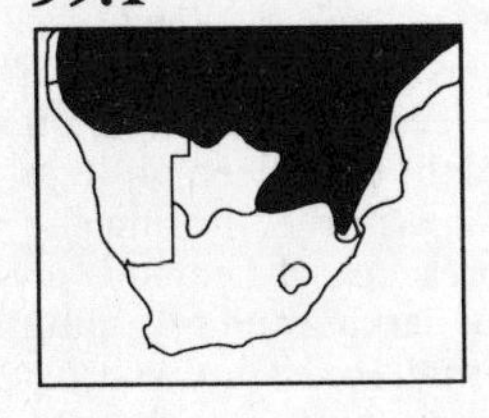

Has huge eyes and large ears. Upper parts are light grey or grey-brown, with under parts lighter. Males: HB 14 cm. T 23 cm. Mass 150 g. Females slightly smaller. **Habitat:** Savanna woodland. Occurs on forest fringes, but not within them. **Habits:** Nocturnal. Aroboreal. Rests by day in family groups of 2 to 7, forages solitarily by night. Has spectacular leaping abilities and can climb amongst the finer outermost twigs. **Food:** Lives on a diet of gum, augmented by insects.

Otolemur crassicaudatus
Thick-tailed Bushbaby
Bosnagaap
59.3

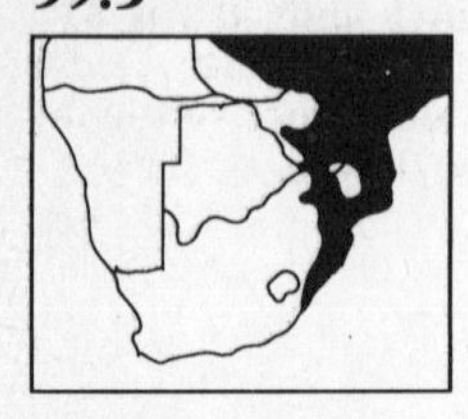

Has huge eyes and large ears. Pale grey in colour tinged with buffy brown, with the under parts lighter. Males: HB 32 cm. T 42 cm. Mass 1,25 kg. Females slightly smaller and lighter. **Habitat:** Forests, thickets and well-developed woodland in the higher-rainfall parts. **Habits:** Nocturnal. Rests during day in groups of 2 to 6, forages solitarily at night. Arboreal. On the ground it moves quadrupedally or hops bipedally. **Food:** Fruit and gum, as well as some insects, reptiles and birds.

Cercopithecus aethiops
Vervet Monkey
Blouaap
60.1

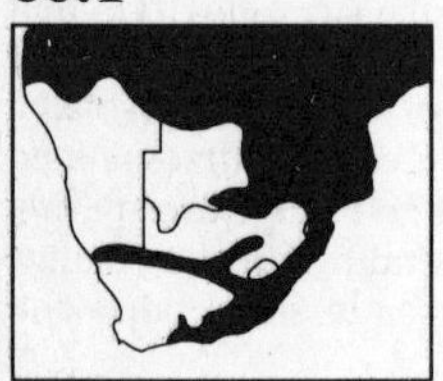

The upper parts are grizzled greyish and under parts whitish. Males: HB 50 cm. T 65 cm. Mass 5,5 kg. Females: HB 45 cm. T 60 cm. Mass 4 kg. **Habitat:** Predominantly savanna woodland. **Habits:** Diurnal. Spends much time in trees searching for wild fruits, but also forages on the ground. It is gregarious, occurring in troops up to 15 or 20. Troops sleep in the higher branches of large trees or in rocky shelters. **Food:** Predominantly vegetarian, it lives on wild fruits, flowers, leaves, seeds and seed pods; also eats insects.

Cercopithecus mitis
Samango Monkey
Samango-aap
60.2

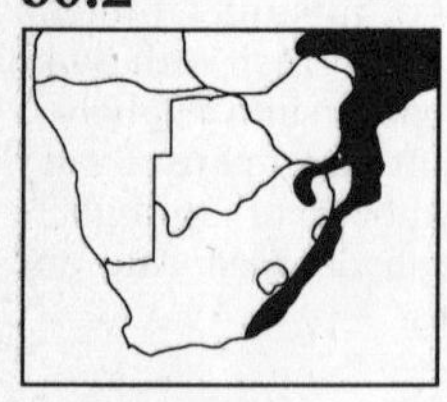

Much darker in colour than Vervet Monkey, and has a dark brown face. Males: HB 60 cm. T 80 cm. Mass 9,3 kg. Females: HB 50 cm. T 70 cm. Mass 4,9 kg. **Habitat:** Closely confined to forest habitat, it seldom strays from it except temporarily when foraging. **Habits:** Diurnal. Rests during hottest time of day. At night it rests in trees, hiding among the foliage. Gregarious, living in troops from 4 to over 30. **Food:** Mainly fruits, dry and green leaves, flowers, pods and shoots, as well as insects.

Papio ursinus
Chacma Baboon
Kaapse Bobbejaan
60.3

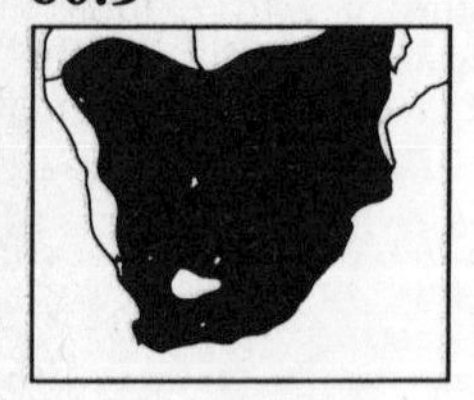

Colour varies with sex and age, as well as geographical areas. Colour may be a dark brown with yellow tinge, or grizzled yellowish-brown. Males: HB 70 cm. T 70 cm. Mass 32 kg. Females: HB 60 cm. T 60 cm. Mass 15,4 kg. **Habitat:** A savanna and montane species, marginal on open grassland. Availability of water essential. **Habits:** Gregarious, troops may number up to about 100 individuals. While the troop feeds, males sitting on vantage points will act as sentinels. At night they sleep on high krantzes or in trees with thick foliage. If disturbed during the night adult males may bark, and the presence of Leopards or other large predators may cause persistent barking and squealing until the danger has passed. They are diurnal. **Food:** Omnivorous, including grasses, seeds, roots, bulbs, leaves, flowers, bark, gum oozing from trees, mushrooms, wild fruits, pods, shoots, insects, spiders, scorpions, ants and slugs. May kill the young of the smaller antelope, as well as hares.

Aonyx capensis
Clawless Otter
Groototter

61.1

The colour of the upper parts varies from light to very dark brown, the under parts lighter in colour. The chest, throat and sides of the face are white. HB 1 m. T 0,5 m. Ht 35 cm. Mass up to 18 kg. **Habitat:** Predominantly aquatic, it occurs in rivers, lakes, swamps and dams. Where it occurs in coastal waters, a supply of fresh water is essential. **Habits:** Predominantly aquatic, it spends much of its active time in water. Generally solitary, but also in pairs and family parties of two adults and up to three young. Predominantly crepuscular, it displays some nocturnal activity in its terrestrial environment, and may occasionally be seen foraging at midday. During the heat of the day it rests in sheltered places. **Food:** Mainly crabs, frogs and fish, and to a lesser extent insects, birds, reptiles, mammals and molluscs.

Lutra maculicollis
Spotted-necked Otter
Kleinotter

61.2

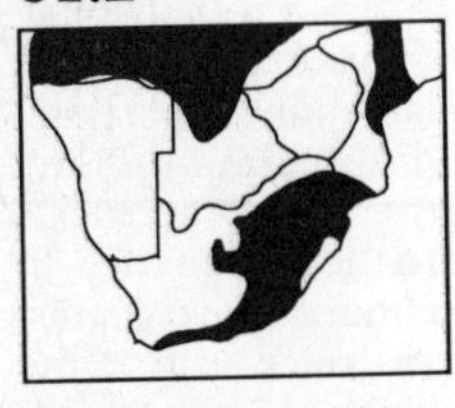

The Spotted-necked Otter is smaller and slimmer than the Clawless Otter. Colour varies from chocolate-brown to a deep, rich reddish-brown. The throat and upper chest are mottled with white or creamy-white. HB 60 cm. T 40 cm. Ht 30 cm. Mass up to 9 kg. **Habitat:** Aquatic, closely confined to large rivers, lakes and swamps which have extensive areas of open water. **Habits:** Usually solitary or family parties of an adult and two young, but schools of up to five or six have been recorded. Apparently crepuscular, but has been seen to move around throughout the day. While it normally creates very little disturbance when diving, it will dive with a resounding splash when alarmed. Lies up in holes in river banks, in rock crevices, or in dense reedbeds. **Food:** Mainly fish, but including crabs, fresh-water molluscs and frogs; also some birds and insects.

Mellivora capensis
Honey Badger
Ratel

62

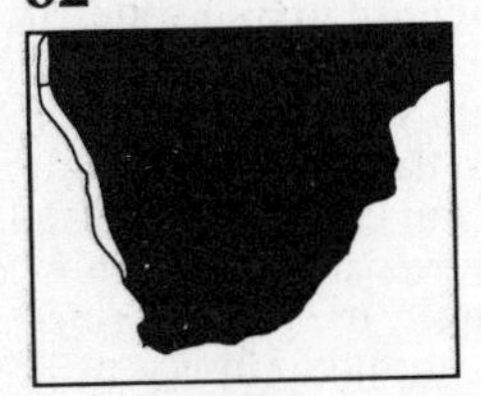

The stocky, short-legged Honey Badger has a broad, light-coloured saddle, which runs from above the eyes to the base of the tail, and contrasts with the black lower parts of the body. Appears to be slightly higher towards the rump. HB 75 cm. T 20 cm. Ht 25–28 cm. Mass 8–16 kg. **Habitat:** Has a wide habitat tolerance, occurring in rocky koppies, scrub sandveld, open grassland, open woodland, riverine woodland, floodplain grassland and at least on the fringes of montane forests. Will use crevices in rocky areas, adapt existing disused holes or dig its own in which to shelter. **Habits:** Predominantly nocturnal, with some diurnal activity where undisturbed. Generally solitary, two or more may hunt together. While normally shy and retiring, it can, without provocation, become extremely aggressive and has a reputation for ferocity and fearlessness. Normally terrestrial, it can climb trees with stout branches to get at beehives. Its thick skin protects it when raiding beehives. **Food:** Omnivorous, including scorpions, spiders, mice, lizards, centipedes, grasshoppers, small birds, snakes, wild berries, fruit, bee larvae and honey. There is good evidence that it follows the Honey Guide (*Indicator indicator*) to a bees' nest.

Poecilogale albinucha
Striped Weasel
Slangmuishond
63

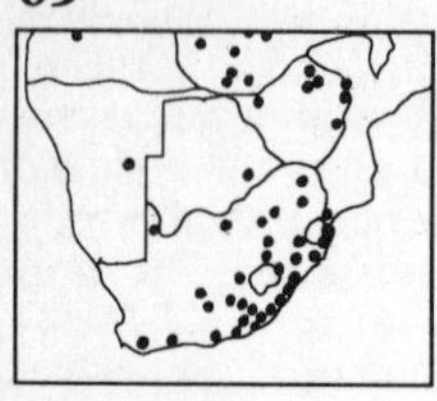

A slender and sinuous little animal with distinct white longitudinal stripes on the jet black fur of the dorsal surface, and a bushy white tail. Smaller than the Striped Polecat, with shorter silky fur and much shorter legs. HB 30 cm. T 15 cm. Ht 5–6 cm. Mass ♂ 260 g, ♀ 170 g. **Habitat:** A savanna species particularly associated with grassland. Uncommon in most parts of its range. **Habits:** Mainly nocturnal, but may be active by day, especially in cool weather. Generally solitary, but also in pairs or family parties. Predominantly terrestrial. A good digger and excavates its own burrows. Relies on its warning coloration for protection, as well as the ejection of a foul-smelling fluid from the anal glands. **Food:** Carnivorous, including warm-blooded prey such as murids.

Ictonyx striatus
Striped Polecat
Stinkmuishond
64

A black animal with distinctive longitudinal white stripes on the sides, and a mainly white tail. Larger than the Striped Weasel, lacking the latter's sinuous body shape and having much longer fur. Its conspicuous coloration serves as a warning of the nauseous ejection from its anal glands. HB 35 cm. T 26 cm. Ht 10–13 cm. Mass ♂ 970 g, ♀ 710 g. **Habitat:** Has a wide habitat tolerance, occurring in open grassland, savanna woodland, thornbush, rocky areas, forest, exotic plantations and along drainage lines in desert. **Habits:** Nocturnal, terrestrial and solitary. Occasionally pairs or female with young are seen. If cornered it is prone to take up aggressive attitudes and turn the hindquarters to the aggressor to eject the pungent excretion from the anal glands. During the day it lies up in holes, hollow trees, thick bush, in the shelter of piles of rocks, or in its own burrows. **Food:** Mainly insects and mice, but will also take reptiles, birds, Amphibia, spiders, scorpions, millipedes and centipedes.

Civettictis civetta
African Civet
Afrikaanse Siwet
65

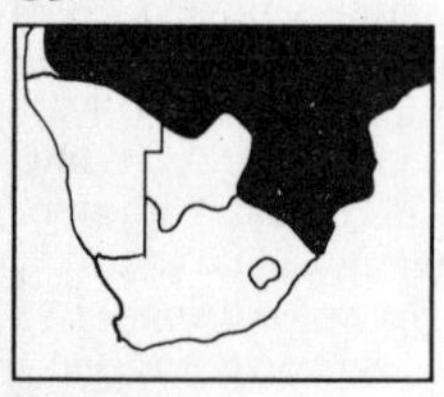

The Civet has a greyish or whitish shaggy coat with black spots on the body, and black stripes on the tail and neck region. The lower parts of the legs are black and the tail is bushy. HB 80 cm. T 46 cm. Ht 35 cm. Mass 12 kg. **Habitat:** Confined to well-watered savanna and forest, not occurring in the dry western and southern areas of the Subregion. Prefers good cover of high grasses, under bush, thickets or reedbeds. **Habits:** Predominantly nocturnal. Terrestrial. Generally solitary, but family parties of an adult and two young have been seen. It moves and feeds silently, and moves off quietly when sensing danger. Its senses of smell and hearing are acute. **Food:** Omnivorous, including insects, wild fruit, murids, reptiles, birds, Amphibia, Myriapoda and carrion.

Nandinia binotata
Tree Civet
Boomsiwet
66

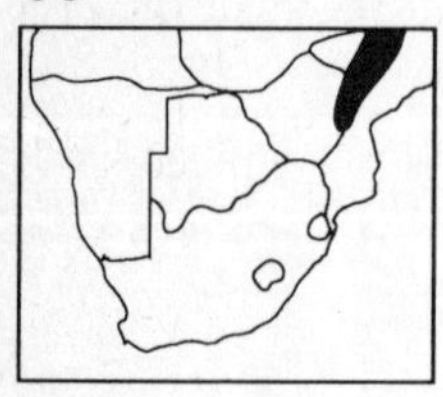

Very similar to a genet, but slightly larger and heavier. The brown woolly fur has faint dorsal spots, the thick tail is ringed, and there is a pair of light spots above the shoulder blades. HB 45 cm. T 45 cm. Ht 22 cm. Mass 2 kg. **Habitat:** Forest. **Habits:** Nocturnal, solitary and predominantly arboreal, but displays some terrestrial activity. By day sleeps in holes in trees or uses the cover of creepers. **Food:** Mainly vegetarian, but also takes carrion and may raid poultry.

Genetta genetta
Small-spotted Genet
Kleinkolmuskejaatkat

67.1

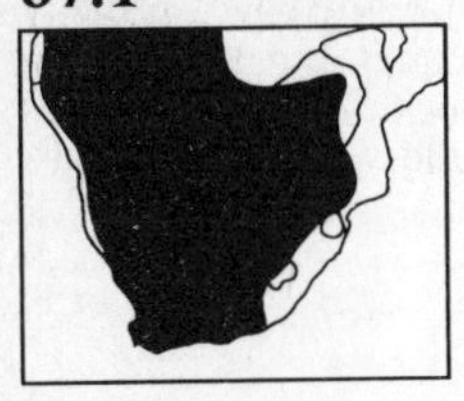

Distinguished from the Large-spotted Genet by a crest of black hair along the back, a longer and coarser coat, more black on the hind-feet, darker body spots, and usually a white-tipped tail, as opposed to the black-tipped tail of the Large-spotted Genet. HB 50 cm. T 45 cm. Ht 15–20 cm. Mass 2 kg. **Habitat:** Utilises the more open areas of savanna woodland, dry grassland or dry vlei areas. Cover, in the form of scrub or underbush, holes in the ground or in trees in which to shelter during day, is essential. **Habits:** Nocturnal. Mainly solitary, but also in pairs. Although mainly terrestrial, it is a good climber. By day it rests up in holes in the ground, hollow logs, holes in trees or the shelter of piles of boulders. **Food:** Small rodents, birds, reptiles, insects, spiders and scorpions.

Genetta tigrina
Large-spotted Genet
Rooikolmuskejaatkat

67.2

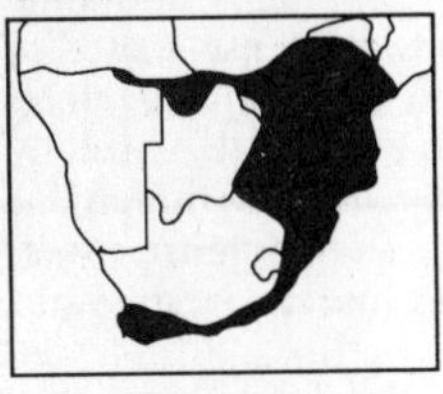

The Large-spotted Genet has a black-tipped tail, as opposed to the white-tipped tail of the Small-spotted Genet, and also lacks the spinal crest. The spots on the body are generally larger than in the Small-spotted Genet. HB 50 cm. T 50 cm. Ht 18 cm. Mass 2 kg. **Habitat:** Particularly associated with well-watered country, usually absent from arid areas and restricted to riverine associations in the drier parts. Cover is essential. **Habits:** Nocturnal. Mainly solitary, but also in pairs. Is a good climber. By day it rests up in holes in trees, in hollow logs, under tree roots, in disused Antbear or Springhare holes, or under piles of boulders. **Food:** Rats, mice, insects, including locusts and beetles, ground birds, scorpions, spiders, reptiles and wild fruit.

Herpestes ichneumon
Large Grey Mongoose
Groot Grysmuishond

68.1

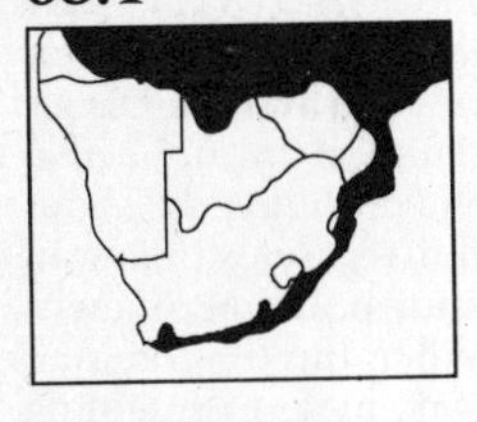

A large grey mongoose with coarse fur, and a long tail which has a distinct black tip. The lower parts of the legs are black. The hair on the flanks and hind-quarters is long, hiding most of the hind-legs. HB 56 cm. T 52 cm. Ht 20 cm. Mass 3,4 kg. **Habitat:** Closely associated with water, occurring in riverine forest and bush or any thick cover not far from water. It frequents reedbeds along rivers, the fringes of lakes, dams and swamps. **Habits:** Mainly diurnal. Pairs and family parties up to 5 may be seen, as well as solitary individuals. Terrestrial, it swims well and will hunt in shallow water. **Food:** Frogs, fish, crabs and murids as well as birds, reptiles and insects.

Galerella sanguinea
Slender Mongoose
Swartkwasmuishond

68.2

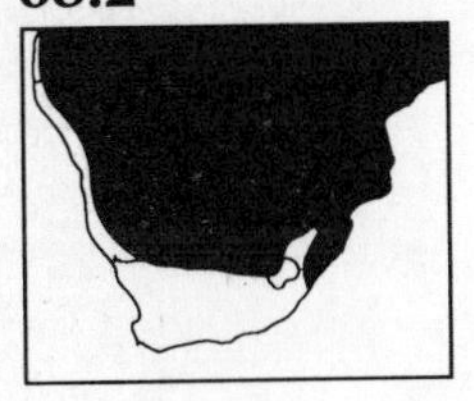

A small slender mongoose with a long tail which has a characteristic black tip. Colour varies, including reddish, yellowish, greyish or dark brown that looks black in the field. HB 30 cm. T 27 cm. Ht 10–12 cm. Mass ♂ 640 g, ♀ 460 g. **Habitat:** Has a wide habitat tolerance, from arid to well-watered areas. Does not occur in desert. Occurs in woodland and on fringes of montane or lowland forest, but not within forest itself. **Habits:** Terrestrial and solitary. A good climber. Predominantly diurnal. It shelters in disused Antbear holes or holes in termitaria. **Food:** Mainly insects, including grasshoppers, termites, beetles and ants, as well as lizards, murids, birds, snakes, frogs, scorpions, centipedes and wild fruits.

Galerella pulverulenta
Small Grey Mongoose
Klein Grysmuishond

68.3

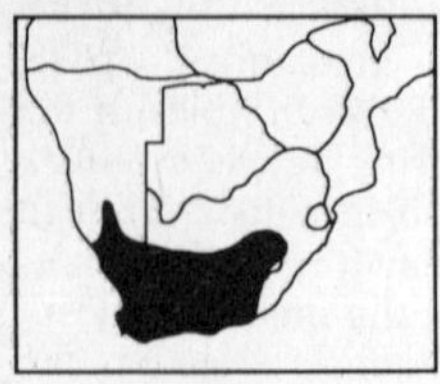

A small, slim mongoose, but heavier and more stoutly built than the Slender Mongoose. General colour is speckled grey. HB 35 cm. T 30 cm. Ht 10–12 cm. Mass ♂ 910 g, ♀ 680 g. **Habitat:** Wide habitat tolerance. Found in open country as well as forest. **Habits:** Diurnal, less active during the heat of the day. Normally solitary, occasionally in pairs. Will climb trees, but is predominantly terrestrial. **Food:** Mainly insects, including grasshoppers and locusts, as well as rats and mice, reptiles and ground birds, their eggs and young; also fruit and carrion.

Atilax paludinosus
Water Mongoose
Kommetjiegatmuishond
(Watermuishond)

68.4

A robust mongoose with a coarse shaggy coat and a tapering tail. General colour varies, usually dark brown, sometimes almost black. HB 55 cm. T 35 cm. Ht 18–20 cm. Mass 3,4 kg. **Habitat:** Associated with well-watered terrain, living in the vicinity of rivers, streams, marshes, swamps, wet vleis, dams and tidal estuaries, where there is cover of reedbeds, or dense stands of semi-aquatic grasses. May penetrate into dry country along rivers. **Habits:** Crepuscular, apparently also nocturnal. Normally solitary, adult females may be accompanied by juveniles. Terrestrial, but is an excellent swimmer. Lies up in dense cover. **Food:** Frogs, crabs, murids, fish, as well as insects, fresh water mussels and vegetable matter.

Helogale parvula
Dwarf Mongoose
Dwergmuishond

68.5

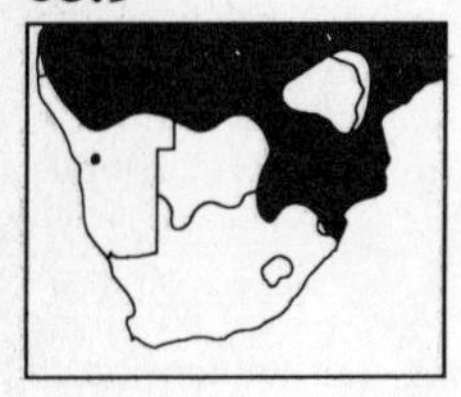

The smallest of the African mongooses. General colour is a uniform speckled brown or reddish, appearing at a distance very dark brown or almost black. HB 22 cm. T 17 cm. Ht 7–8 cm. Mass 270 g. **Habitat:** A savanna species associated with dry open woodland and grassland. Uses holes in termitaria as shelter. Does not occur in semi-desert, desert or in forest. **Habits:** Diurnal, active only when sun is well up in the morning and retiring early, about 16h00. Terrestrial and reluctant to climb. Gregarious, living in troops of 8–10, but troops of 20–30 are known. Has permanent residences, usually in termite mounds. When the troop moves out to forage, members scatter over a wide area, maintaining contact by vocalising. **Food:** Mainly insectivorous, diet includes termites, locusts, snails, scorpions, centipedes, earthworms, reptiles and the eggs of ground birds and of snakes.

Mungos mungo
Banded Mongoose
Gebande Muishond

68.6

A small mongoose with a coarse and wiry coat, about a dozen black transverse bands on the back and a short tapering tail. Ground colour is brownish grey. HB 35 cm. T 24 cm. Ht 18–20 cm. Mass 1,4 kg. **Habitat:** Wide habitat tolerance, but generally absent from forest, desert or semi-desert areas. Occurs in savanna, thickets, scrub thickets and in dry forest. **Habits:** Gregarious, living in troops which vary in size from a few to over 30. May scatter when foraging, maintaining contact by a continuous twittering. Diurnal; nocturnal activity has not been recorded. Dens are in disused Antbear holes, erosion gullies and termite mounds. Terrestrial, but will climb trees under stress. **Food:** Insects, grubs, myriapods, snails, small reptiles, the eggs and young of ground-nesting birds, wild fruits, as well as scorpions and spiders.

Ichneumia albicauda
White-tailed Mongoose
Witstertmuishond

68.7

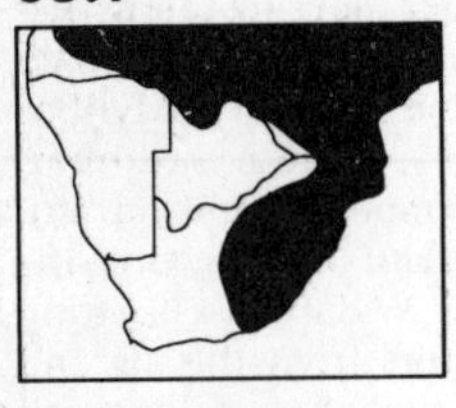

A large, shaggy-coated mongoose, distinguished by longish legs and a tail that is white for about four-fifths of its length. The general colour is a grizzled darkish grey, with under parts of the limbs dark. HB 68 cm. T 42 cm. Ht 22–25 cm. Mass 4,5 kg. **Habitat:** Associated with savanna woodland in well-watered areas. Uses fringes of evergreen forest, but is not found within forest itself. Does not occur in desert or semi-desert. **Habits:** Nocturnal. Terrestrial, it cannot climb trees. Normally solitary, also in pairs or family parties. By day it lies up in disused Antbear or Springhare holes. **Food:** Mainly insects, as well as frogs, murids, reptiles, millipedes, spiders, scorpions, and vegetable matter.

Rhynchogale melleri
Meller's Mongoose
Meller se Muishond

68.8

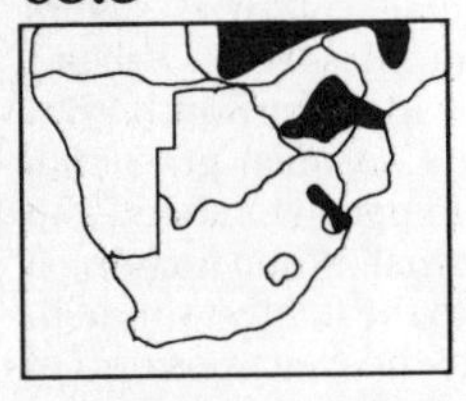

A medium-sized mongoose with long coarse fur. General colour a grizzled light brown. The tail may be black, brown or white. HB 47 cm. T 36 cm. Ht 15–18 cm. Mass 3 kg. **Habitat:** A savanna species particularly associated with open woodland or grassland with termitaria. **Habits:** Nocturnal. Solitary, or female with young. Terrestrial. **Food:** Termites, beetles, grasshoppers, small rodents, lizards and wild fruits.

Paracynictis selousi
Selous' Mongoose
Kleinwitstertmuishond

68.10

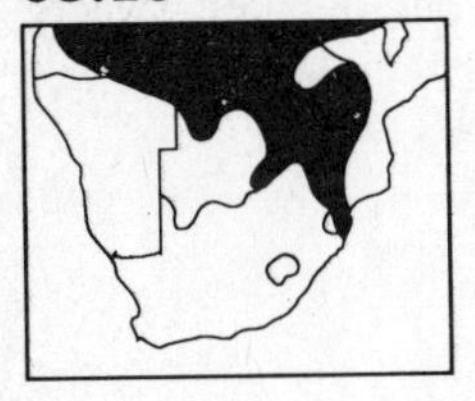

A small but long and slender mongoose with long soft fur. Tail is long and bushy with a white tip. Colour is an overall white-grey; legs dark. It is smaller than the White-tailed Mongoose, and the tail is white only for a short section towards the tip. HB 40 cm. T 38 cm. Ht 15–18 cm. Mass 1,7 kg. **Habitat:** A savanna species associated with more open areas of scrub, woodland, floodplain or grassland. Does not occur in desert, semi-desert or forest. **Habits:** Nocturnal. Terrestrial, with no arboreal tendencies. Normally solitary, also in pairs. Lies up in burrows during the day. Locates food by smell. **Food:** Mainly insects, but also spiders, scorpions, centipedes, rats, mice, small birds, eggs of birds and reptiles, lizards and snakes.

Cynictis penicillata
Yellow Mongoose
Witkwasmuishond

68.11

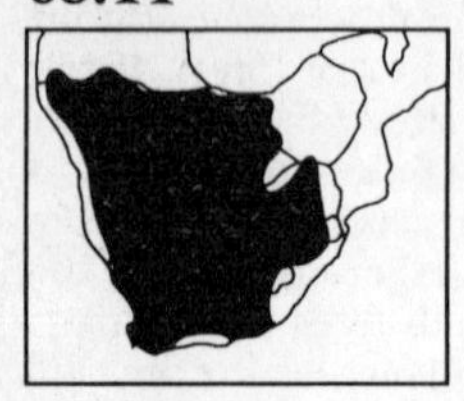

A small mongoose with a short pointed muzzle. The colour varies from a richly coloured tawny-yellowish in the southern parts of the range to a grizzled greyish in the northern parts. The former have long white-tipped tails and long-haired coats, the latter have shorter tails with no white tip and the body is covered with short hair. Botswana: HB 30 cm. T 21 cm. Mass 590 g. Orange Free State: HB 34 cm. T 24 cm. Mass 830 g. Ht 15–18 cm. **Habitat:** Associated with open country, generally of the South West Arid Zone, but also extending into savanna. Does not occur in desert, forest or areas of thick bush. **Habits:** Predominantly diurnal, with some nocturnal activity. Gregarious, living in colonies of up to 20 or more, in warrens. Often shares warrens with Ground Squirrel (*Xerus inauris*) and Suricate (*Suricata suricatta*). It is terrestrial, with no arboreal tendencies. **Food:** Predominantly insectivorous, but also takes mice, small birds, reptiles, scorpions, centipedes, spiders and frogs.

Suricata suricatta
Suricate
Stokstertmeerkat

68.12

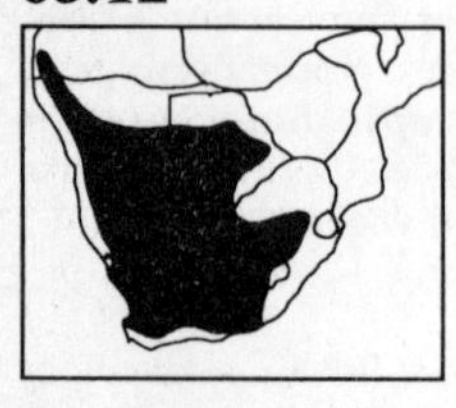

Small and stockily built. Conspicuous black rings around the eyes. General colour is light grizzled fawn with dark brown transverse bands on the back, and a slender tapering tail with a black tip. HB 28 cm. T 22 cm. Ht 15 cm. Mass 730 g. **Habitat:** Occurs throughout the South West Arid Zone, also in the fynbos of the southern Cape Province and the southern savannas. Absent from desert and forest, and avoids mountainous terrain. Has a preference for open, arid country. **Habits:** Diurnal. Gregarious, living in colonies of up to 30 in warrens. Will sun itself, sitting on its haunches. Always alert for predators from the air and ground, and when the alarm call is given the whole colony will dive for their burrows. **Food:** Mainly insects, as well as scorpions, spiders, small reptiles and myriapods.

Otocyon megalotis
Bat-eared Fox
Bakoorvos

69.1

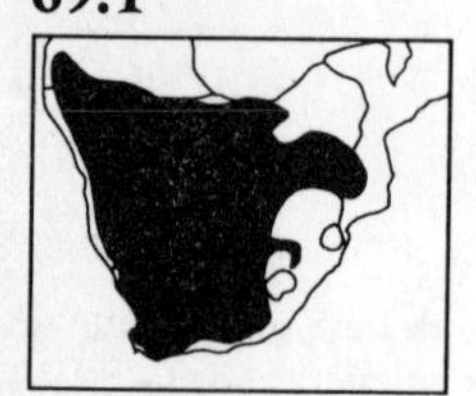

Looks like a small jackal. Has large black-edged ears, blackish legs and a bushy, black-tipped tail. The overall colour is silvery buffy-grey. HB 54 cm. T 30 cm. Ht 30 cm. Mass 4 kg. **Habitat:** Particularly associated with open country within the South West Arid and southern savanna zones. Occurs on open grassland, with a preference for areas of short grass, in open woodland and karoo scrub. **Habits:** Both diurnal and nocturnal. When foraging it moves around apparently aimlessly, alert to the faintest sound in the grass. If it detects a sound it will locate the exact position of subterranean noise among grass roots with its ears close to the ground. The two sets of front claws dig in exactly the same hole in a quick and effective action, making a narrow and deep hole. Pairs or family parties, consisting of two adults and up to four young, are often seen. **Food:** Mainly insects, in particular the Harvester Termite, also scorpions, murids, reptiles, spiders, millipedes, centipedes and wild fruit.

Vulpes chama
Cape Fox
Silwervos
69.2

A small fox, with large pointed ears and a short, pointed muzzle. The upper parts of the body look silvery grey, the under parts pale buffy. The bushy tail is pale fawn with a black tip. HB 55 cm. T 36 cm. Ht 35 cm. Mass 3 kg. **Habitat:** Associated with open country, open grassland, grassland with scattered thickets, or semi-desert scrub, penetrating marginally into open dry woodland and into the fynbos of the southwestern Cape Province. **Habits:** Predominantly nocturnal. By day it lies up in holes or in the cover of stands of tall grass. Generally solitary. The foxes only have contact with each other at the time of mating, and females maintain contact with young until they are ready to disperse. **Food:** Mainly mice and insects; also small mammals, scorpions, spiders, centipedes, birds and eggs, reptiles, carrion, wild fruit and green grass.

Canis mesomelas
Black-backed Jackal
Rooijakkals
70.1

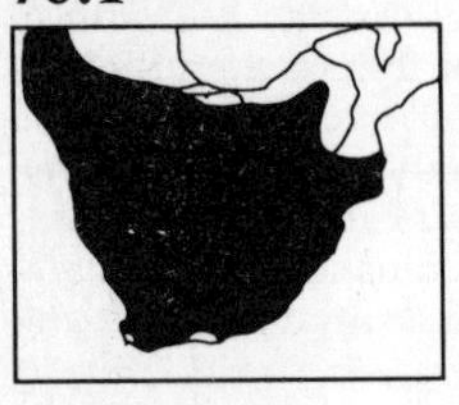

Characterised by reddish body colour, black saddle on the back, and bushy, black-tipped tail. HB 72 cm. T 33 cm. Ht 38 cm. Mass ♂ 8 kg, ♀ 7 kg. **Habitat:** Wide habitat tolerance, occurring in the South West Arid and southern savanna zones. Associated with open terrain. **Habits:** Both diurnal and nocturnal. Normally occurs in pairs or solitary, as well as family parties consisting of parents with 3 to 4 young. Terrestrial. Senses are acute, particularly sense of smell. Will follow up the downwind scent of a dead animal from a distance of well over a kilometre. Rests in holes, in rock crevices or piles of boulders. **Food:** Mainly carrion, but also small mammals, such as rats and mice, insects, vegetable matter, birds, reptiles, sun spiders, scorpions, centipedes and green grass.

Canis adustus
Side-striped Jackal
Witkwasjakkals
70.2

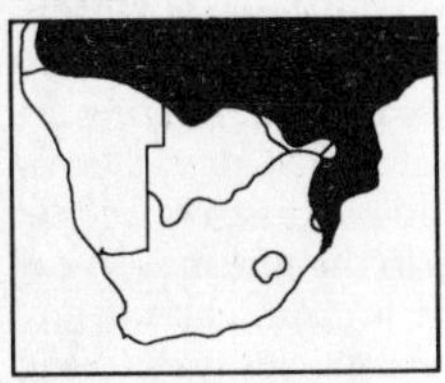

A large jackal with an overall grey or greyish-buff colour. Has a faint black-and-white stripe along the side and a white-tipped tail. HB 72 cm. T 36 cm. Ht 38 cm. Mass 9 kg. **Habitat:** Avoids open savanna grassland, favouring more thickly wooded country, but does not occur in forest. Associated with well-watered habitat. **Habits:** Nocturnal. By day it lies up in holes in the ground or in piles of boulders. Normally solitary, pairs or family parties consisting of a female with young. **Food:** Mainly vegetable matter, including wild fruits and agricultural crops; also small mammals, predominantly rats and mice, insects, carrion, birds and reptiles.

Lycaon pictus
Wild Dog
Wildehond
71

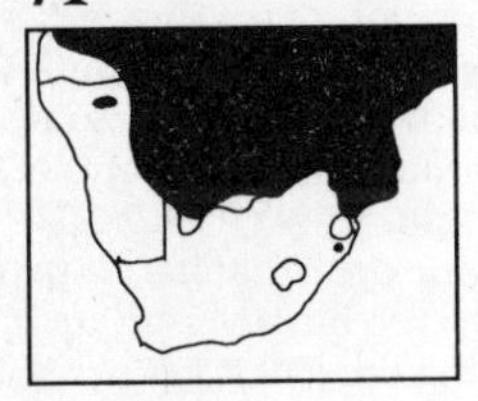

Dog-like in appearance, its body is blotched with black, yellow and white. It has large rounded ears and white-tipped bushy tail. HB 80–108 cm. T 30–40 cm. Ht 70–75 cm. Mass 18–28 kg. **Habitat:** Associated with open plains and open savanna woodland. Utilises open country because it relies on sight for hunting. **Habits:** Adapted to living in packs of about 10–15 individuals, up to 50. Hunts by sight, normally in the early morning and late evening. They are coursers, the pack moving on a broad front up to the prey in open country, breaking into a run when the prey runs off. The prey is singled out by the pack leader and relentlessly pursued, sometimes for several kilometres at sustained speeds of up to 48 km/hour. Generally active only during daylight hours. Lies up during the heat of the day. **Food:** Mainly smaller to medium-sized antelope; also hares and the young of the larger bovids.

Proteles cristatus
Aardwolf
Aardwolf
72

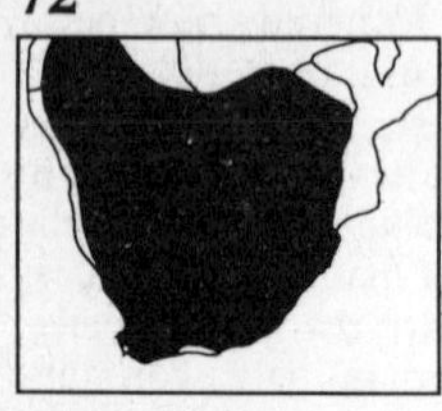

About the size of a jackal, but has the shape of a hyaena, with shoulders sloping down to the back legs. Has a thick-haired mane on the back and a bushy black-tipped tail. General colour is yellowish brown or buff, with vertical stripes on the body. HB 68 cm. T 25 cm. Ht 50 cm. Mass 9 kg. **Habitat:** Wide habitat tolerance. Occurs in South West Arid and southern savanna zones, in open karroid associations, grassland and scrub, and open savanna woodlands. Does not occur in desert or forest. Dependent on the availability of various species of termites. **Habits:** Predominantly nocturnal. By day it lies up in burrows. Normally solitary, but sometimes in pairs or family parties. It is terrestrial. **Food:** Mainly termites, as well as other insects, spiders, and millipedes.

Hyaena brunnea
Brown Hyaena
Strandjut
73.1

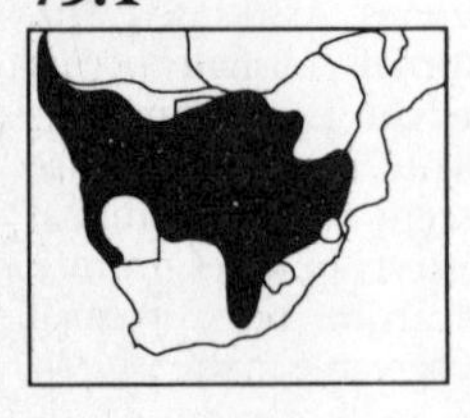

In profile higher at the shoulders than at the rump, it has a long, shaggy brown coat, a lighter-coloured mantle and a long bushy tail. The Brown Hyaena is smaller than the Spotted Hyaena. HB 120 cm. T 20 cm. Ht 80 cm. Mass ♂ 47 kg, ♀ 42 kg. **Habitat:** Occurs in desert, in semi-desert, open scrub and open woodland savanna. **Habits:** Predominantly nocturnal. By day it lies up in holes in the ground or in the shelter of bushes or clumps of tall grass. Lives in groups that occupy fixed territories and will scavenge communally but forage solitary. Senses are well developed, especially that of scent, since it can detect carcasses over considerable distances. **Food:** Predominantly a scavenger, but also hunts small mammals, birds, reptiles, and eats fruit and insects.

Crocuta crocuta
Spotted Hyaena
Gevlekte Hiëna
73.2

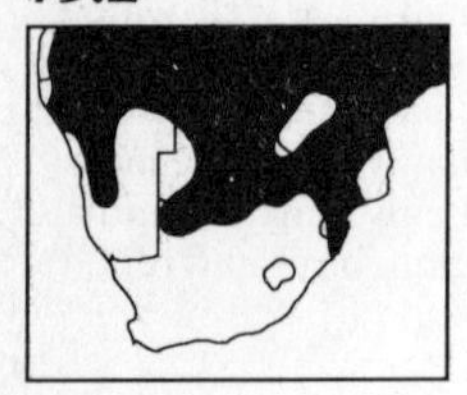

Characteristically the shoulders are heavier and stand much higher than the hind-quarters. The dull yellowish coat is marked by irregular dark spots. HB 136 cm. T 26 cm. Ht 80 cm. Mass ♂ 60 kg, ♀ 70 kg. **Habitat:** A savanna species associated with open plains, open woodland and semi-desert scrub. Absent from forest, except marginally. **Habits:** Social organisation is based on a matriarchal system of clans. Senses of sight, smell and hearing are acute. Predominantly nocturnal, but also active by day. During the hotter parts of the day it rests in the shade of bushes, trees or among piles of boulders, or it will use holes in the ground. Its call is a characteristic series of long-drawn-out *whoops*, each beginning low and rising high. Also grunts, groans, giggles, yells. **Food:** Predominantly large- or medium-sized ungulates, but will hunt or scavenge a wide range of other prey, including Springhare, birds, fish, reptiles, crabs, snails and termites as well as fruit. Locates a carcass by observing vultures circling and descending on it.

Felis nigripes
Small Spotted Cat
Klein Gekolde Kat
74.1

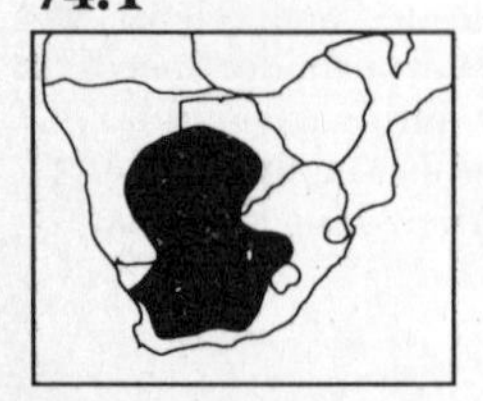

The smallest cat occurring in the Subregion. The general colour is tawny, marked with large black spots and transverse stripes on the shoulder. It has three black bands around the legs, and a bushy tail with a black tip. HB 40 cm. T 18 cm. Mass ♂ 1,6 kg, ♀ 1,1 kg. **Habitat:** Particularly associated with open habitat with some cover, such as stands of tall grass or scrub bush. Associated with arid country. **Habits:** Nocturnal. During the day it lies up in disused Springhare or Antbear holes or holes in termite mounds. Solitary. Capable of climbing trees, but is almost exclusively a terrestrial feeder. **Food:** Mainly murids, as well as spiders, insects, reptiles and birds.

Felis lybica
African Wild Cat
Vaalboskat
74.2

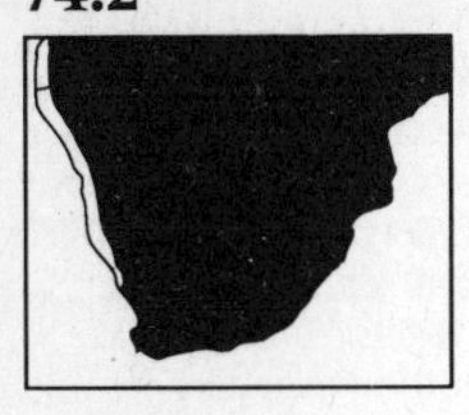

Looks very much like the Domestic Cat, but slightly larger. General colour from greyish to buffish or ochraceous, with dark spots and stripes. Males: HB 60 cm. T 30 cm. Mass 5 kg. Females: HB 52 cm. T 30 cm. Mass 4 kg. **Habitat:** Wide habitat tolerance. **Habits:** Nocturnal. By day it lies up in the cover of underbush, reedbeds, stands of tall grass or rocky hillsides. Solitary, except when several males accompany female in oestrus. Terrestrial, but is adept at climbing trees under stress or when hunting. **Food:** Mainly rats and mice, as well as birds, reptiles, insects, spiders and small mammals such as hares, Springhare and the young of small antelope.

Felis serval
Serval
Tierboskat
74.4

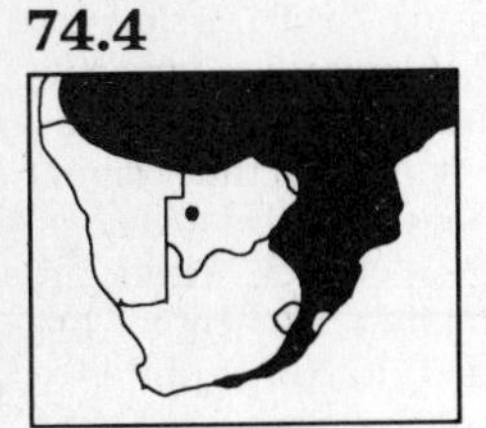

A slender cat with long legs, small head, large ears and spotted and barred coat. The general colour is yellow with black spots. HB 80 cm. T 30 cm. Ht 55–60 cm. Mass ♂ 11,1 kg, ♀ 9,7 kg. **Habitat:** Confined to areas where there is permanent water, in the higher rainfall areas. Does not occur in desert or semi-desert, but can penetrate arid country along intrusions of better-watered terrain. **Habits:** Predominantly nocturnal. Normally solitary, also in pairs or females with young. When disturbed it takes cover in the nearest stand of tall grass or reedbeds or takes to hillsides where there is good cover of underbush. Will climb trees under stress. **Food:** Mainly rats and mice, as well as other small mammals, birds, reptiles, insects and sun spiders.

Felis caracal
Caracal
Rooikat
74.5

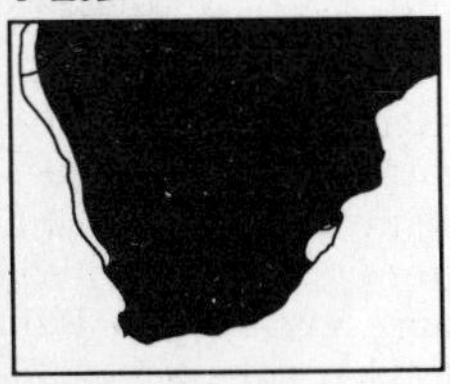

The Caracal is stockier than the Serval, with shorter limbs, a short bushy tail and characteristic tufts on the ears. Overall colour is a reddish tan. HB 80 cm. T 30 cm. Ht 40–45 cm. Mass ♂ 14 kg, ♀ 11 kg. **Habitat:** Associated with open savanna woodland, open grassland and open vleis, semi-desert, karroid and savanna areas. Absent from true desert and forest. **Habits:** Predominantly nocturnal. By day it is able to conceal itself in the most meagre cover. Solitary, only associating for mating and that for short periods. Normally terrestrial, it is adept at climbing trees. **Food:** Predominantly small- and medium-sized prey, including small mammals, the young of larger antelopes, birds and reptiles.

Acinonyx jubatus
Cheetah
Jagluiperd
75.1

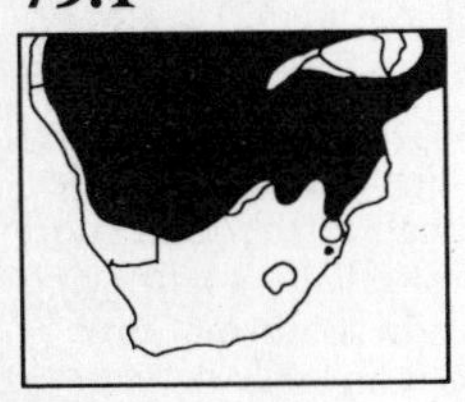

The Cheetah is a much longer, more slender animal than the Leopard, with longer legs, a much smaller head, and characteristic black 'tear marks' from the eyes to the mouth. The spots are much smaller and more rounded than the rosettes of the Leopard. HB 130 cm. T 70 cm. Ht 85 cm. Mass ♂ 54 kg, ♀ 43 kg. **Habitat:** Open plains and the more open areas within savanna woodland, as well as the fringes of desert. **Habits:** Predominantly diurnal. During the hottest hours it lies up in the shade. Occurs in pairs or family parties of 3 or 4, and to a lesser extent solitary. Terrestrial, the Cheetah is ill-adapted to climbing, and averse to swimming. **Food:** Mainly medium-sized or small bovids or the young of larger bovids, as well as terrestrial birds and small mammals such as hares and Porcupines. When two or more hunt together they may kill larger ungulates. In open country it simply walks up to the prey, pausing motionless when the prey shows anxiety. In woodland or scrub country it will use cover in stalking. Gives chase when the prey takes fright and runs off, maintaining maximum speed for about 300–400 m.

Panthera pardus
Leopard
Luiperd
75.2

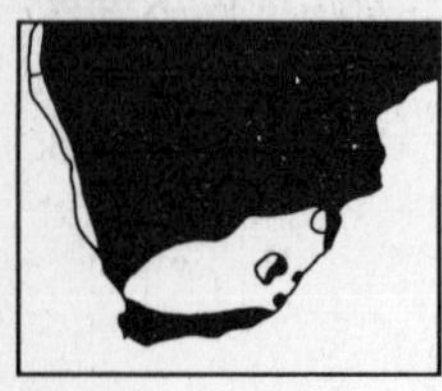

The Leopard is more solid-bodied, with shorter, stockier legs, and a larger head than the Cheetah. Golden yellow in colour, with distinct black, light-centred rosettes. Males: HB 130 cm. T 80 cm. Ht 50–70 cm. Mass 50–60 kg. Females: HB 110 cm. T 75 cm. Ht 45–60 cm. Mass 30–40 kg. The Cape Mountain Leopard is usually smaller than that of the bushveld, with mass of 30–40 kg for males and 20–25 kg for females. **Habitat:** Wide habitat tolerance. Occurs in areas of rocky koppies, rocky hills, mountain ranges, bushveld savanna and forest, as well as semi-desert. Will penetrate desert along avenues of watercourses. Cover to lie up in by day and to hunt from is essential. **Habits:** Solitary, except during mating season or when female is accompanied by young. Mainly nocturnal. During the hotter hours of the day it lies up in dense cover, the shade of rocks or caves. Predominantly terrestrial, it is a good tree-climber and negotiates steep rocky areas. The Leopard is a secretive and silent animal. Its senses are well developed, particularly sight and hearing. **Food:** Dassies, rats and mice, hares, small to medium-sized ungulates, sometimes larger mammals such as Kudu and hartebeest, birds, baboons, snakes, lizards, insects, scorpions and some of the smaller carnivores, such as jackals and Domestic Dogs. Will also scavenge. When natural prey is scarce it kills domestic stock. There are many records of healthy Leopards turning into man-eaters, but such behaviour is atypical of the Leopard in the Subregion.

Panthera leo
Lion
Leeu
75.3

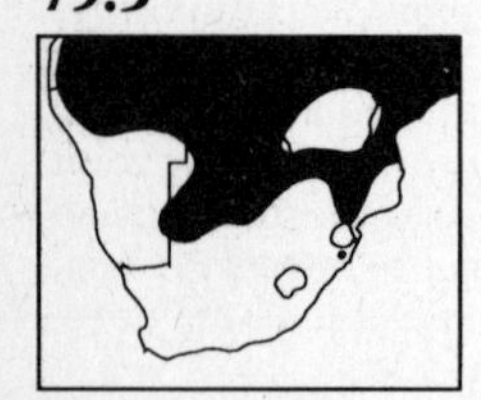

The largest of the African carnivores, the colour of the body is sandy or tawny, the males carrying a mane of thick hair around the neck. HB 145–200 cm. T 67–102 cm. Ht 75–125 cm. Mass 120–200 kg. Males larger and heavier than females. **Habitat:** Wide habitat tolerance, but does not occur in forest. Will penetrate desert along avenues of watercourses, and is common in semi-desert areas. **Habits:** Predominantly nocturnal and active around sunrise and towards sunset, but often also active during daylight hours. During the heat of the day it lies up in the shade. Terrestrial, but is a good climber. If suddenly disturbed during resting periods it can quickly become aggressive. The Lion lives and hunts in prides which may number from a few individuals to 30 or more. **Food:** Mainly medium-sized to large ungulates, but will kill a wide range of mammals, from mice to Buffalo. Also birds, reptiles and even insects. Has been recorded to kill and eat Spotted Hyaenas, Leopards, Cheetahs, jackals, civets, Honey Badgers, Caracal and even Crocodiles. Is known to appropriate the kills of other carnivores such as hyaenas and Leopards, and will take carrion. Is an expert stalker and will make use of the barest cover to get close to its prey. If the prey shows signs of anxiety it will freeze motionless until the former relaxes and continues feeding before moving forward again. A careful stalk is followed by a quick rush to seize the animal. In the final short sprint a Lion may cover 100 m in 6 seconds. Remains silent while hunting. Prey is killed by strangulation. Where normal prey is absent, will take cattle and small stock. Can also become a man-eater.

Orycteropus afer
Antbear (Aardvark)
Erdvark
76

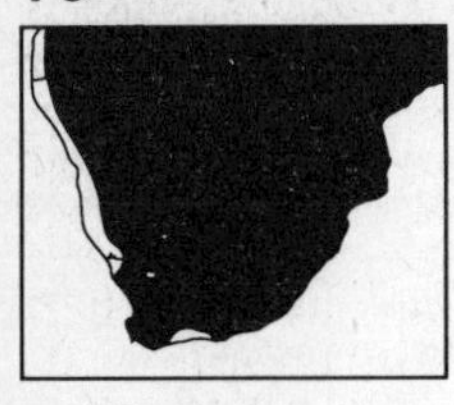

The Antbear has a humped back, long pig-like snout, long donkey-like ears and a thick tapering tail. Its body is pale yellowish-grey and sparsely covered with hairs. HB 105 cm. T 55 cm. Ht 60–65 cm. Mass 53 kg. **Habitat:** Wide habitat tolerance, including open woodland, scrub and grassland. Particularly associated with heavily utilised grassland where there are termite populations. **Habits:** Almost exclusively nocturnal. By day it hides in burrows which it excavates. Solitary, except female with young or pairs at time of mating. Senses of smell and hearing well developed, but sight poor. **Food:** Mainly termites and ants, including the eggs and larvae. Termites are taken chiefly during the wet season and ants during the dry. Also eats beetle larvae, locusts and wild cucumber seeds. When a nest is located, it digs until the nose and mouth can be inserted. Inserts its long, slimy tongue into the tunnels, withdrawing it into the mouth covered with ants.

Loxodonta africana
Elephant
Olifant
77

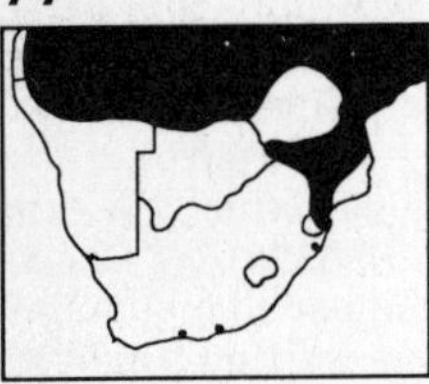

The Elephant is the largest living land mammal. The trunk is used for gathering food, sucking up water, chastening youngsters, smelling, trumpeting, breathing, and as a weapon. Apart from being organs of hearing, the ears have many blood vessels which facilitate heat loss when the Elephant flaps its ears to cool off. The tusks are used as weapons, as well as for digging and prising bark off trees. Females can be distinguished from males by the much more pronounced forehead. Males: Ht 3–3,5 m, up to 4 m. Mass 4 500–5 000 kg, up to 6 000 kg. Females: Ht 2,4–2,8 m, up to 3 m. Mass 2 200–2 500 kg, up to 3 000 kg. **Habitat:** Wide habitat tolerance. Occurs in woodland, savanna, plains, semi-desert and on the fringes of the coastal Namib Desert. Will wander down river courses in the Namib. Essential habitat requirements include clean, sweet water, a plentiful supply of food in the form of palatable grasses and browse plants, and some shade in which to shelter during the hottest hours of the day. **Habits:** Gregarious. Family groups consist of an adult female with her offspring or a number of closely related females with their offspring. Family groups may combine to form herds. Bulls may join male herds or remain on their own. Very old bulls often live solitary lives. Large associations numbering hundreds of individuals of all age classes and sexes may be found. Elephant society is matriarchal. Both diurnal and nocturnal. During the hottest hours of the day the Elephant stands under shady trees. Whenever it seeks water, it will bathe, and is partial to wallowing in mud or dusting itself to protect its skin from the hot sun. Can swim, or will walk on the bottom with the tip of the trunk above the water. Sense of smell very keen, but eyesight and hearing are not very good. Vocalisation consists of trumpeting and screaming when upset, and Elephants maintain contact with a deep rumbling. A feeding herd can be very noisy as they break branches down. When disturbed, the Elephant remains dead quiet and will move off without a sound. **Food:** Browse and graze, utilising a wide range of species; also eats the bark of certain trees. Partial to sweet, clean water.

Equus burchelli
Burchell's Zebra
Bontsebra (Bontkwagga)

78.1

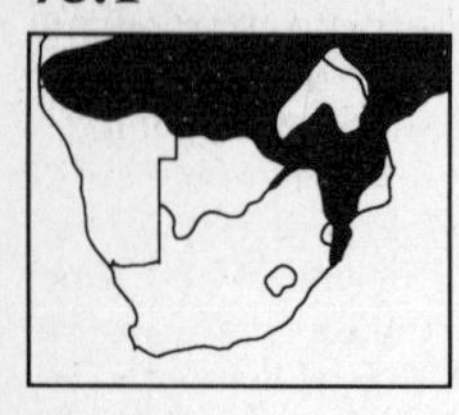

Burchell's Zebra may be distinguished from the two mountain zebras (*E. zebra*) by the yellowish or greyish shadow stripes between the black on the hind-quarters and the lack of the 'gridiron' pattern on top of the hind-quarters. Shadow stripes vary from distinct to missing altogether. The black stripes of individuals are never exactly alike. HB 230 cm. T 45 cm. Ht 136 cm. Mass 320 kg. **Habitat:** A savanna species, partial to open areas of woodland, open scrub and grassland. **Habits:** Generally active throughout the day. At night it rests for periods with short grazing spells in between. Gregarious, living in small family groups consisting of a stallion and one or more mares and their foals. Surplus stallions form bachelor groups or remain solitary. When a family group is attacked by predators, the family group stallion will take up the rear position and defend his group by kicking or biting. Senses of sight, smell and hearing are acute, and for this reason other species, such as wildebeest, are frequently associated with them, since the Zebra will alert them to signs of danger. **Food:** Predominantly a grazer, but will occasionally browse and feed on herbs.

Equus zebra zebra
Cape Mountain Zebra
Kaapse Bergsebra (Kaapse Bergkwagga)

78.2

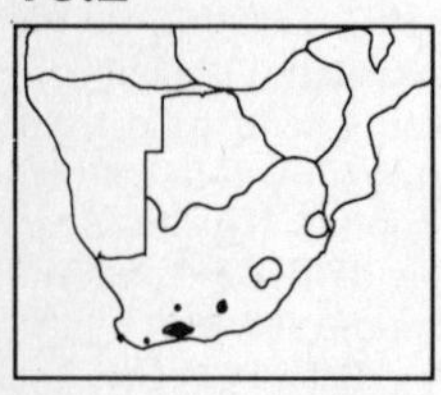

The Mountain Zebra differs from Burchell's Zebra in that the black body stripes do not continue onto the white under parts, and have no shadow stripes between them; moreover, on the rump the black markings form a 'gridiron' pattern which is a characteristic feature. The Mountain Zebra is smaller in body size than Burchell's Zebra, and the Cape Mountain Zebra is in turn smaller than Hartmann's Mountain Zebra. HB 225 cm. T 42 cm. Ht 125 cm. Mass 250 kg. **Habitat:** Closely confined to mountainous areas that offer the required types of grazing and supply of water, as well as shelter in the form of kloofs and ridges. **Habits:** Predominantly diurnal. Dusting is regularly practised by rolling in the sand. Gregarious; breeding herds consist of a stallion and his mares with their foals, ranging in numbers from 2 to 13. When danger threatens, the stallion maintains a defensive position at the rear. Uses kloofs and caves for shelter. **Food:** Predominantly a grazer, but will also browse.

Equus zebra hartmannae
Hartmann's Mountain Zebra
Hartmann se Bergsebra (Bergkwagga)

78.3

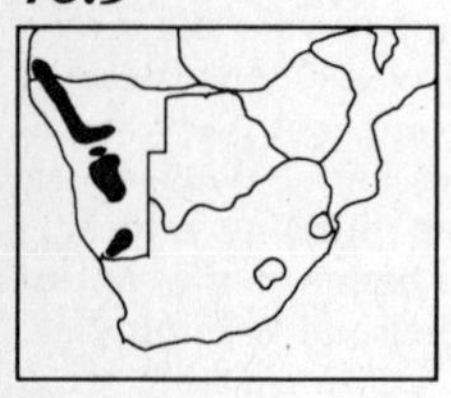

Hartmann's Mountain Zebra is slightly larger than the Cape Mountain Zebra. Ht 150 cm. Mass ♂ 300 kg, ♀ 276 kg. **Habitat:** Prefers the ecotone of mountainous areas and flats. Uses kloofs and krantzes as shelters from cold winds. **Habits:** Diurnal. Gregarious. Family groups consist of a stallion with his mares and their foals. Stallion groups may temporarily attach themselves to family groups. Solitary stallions are occasionally found. During the hotter hours of summer days it rests in the shade of trees. Dust bathes regularly throughout the day. **Food:** Mainly a grazer.

Ceratotherium simum
Square-lipped Rhinoceros (White Rhinoceros)
Witrenoster

79.1

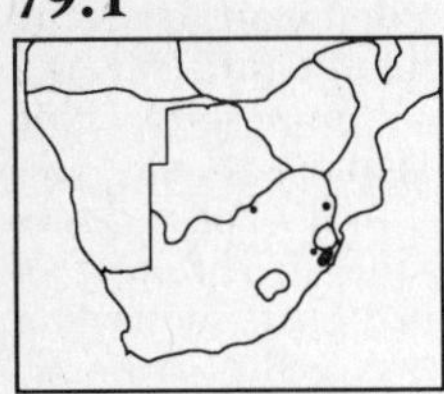

Characteristic features include the square upper lip and prominent hump above the shoulders. It is larger in size and has a longer head than the Hook-lipped Rhinoceros. HB 380 cm. T 100 cm. Ht 180 cm. Mass ♂ 2 000–2 300 kg, ♀ 1 400–1 600 kg. **Habitat:** Areas of short grass, adequate thick bush cover and relatively flat terrain. Wooded grasslands. **Habits:** Grazing and resting at intervals of a few hours, at night, morning, late afternoon and evening. During the heat of the day it rests in the shade of trees. Occurs in small groups consisting of a dominant bull, subordinate bulls, cows and their offspring. Territorial bulls occupy clearly defined territories, which are defended against bulls from neighbouring territories. Has poor sight but acute senses of smell and hearing. **Food:** A grazer, with a preference for short grass.

Diceros bicornis
Hooked-lipped Rhinoceros (Black Rhinoceros)
Swartrenoster

79.2

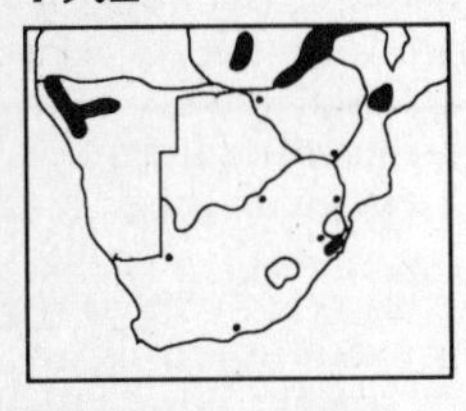

It is distinguished from the Square-lipped Rhinoceros by its smaller size, its pointed upper lip and its smaller head. HB 360 cm. T 70 cm. Ht 160 cm. Mass up to 1 000 kg. **Habitat:** Occurs in wide range of habitats from forest to savanna woodland and scrub, but not in open plains. **Habits:** Solitary, or female with calf. Associations of an adult male with female or a number of individuals of all ages, are transitory. Rhinos are not territorial, and tend to avoid each other. Active during the early morning or late afternoon. During the heat of the day it rests in the shade. Sight is poor, but senses of hearing and smell are acute. **Food:** Mainly browse, but small quantities of grass are taken during the wet season.

Hippopotamus amphibius
Hippopotamus
Seekoei

80

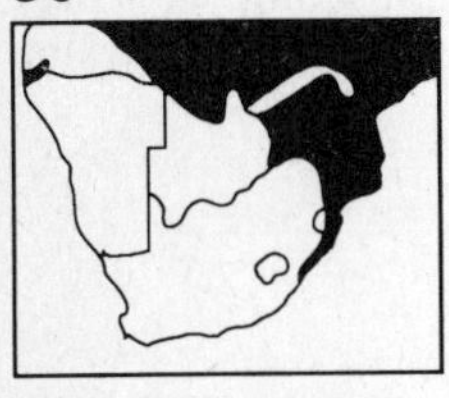

A special adaptation for its aquatic life is the positioning of the eyes, nose and ears, which all protrude from the water when the rest of the animal is submerged. HB 370 cm. Ht 150 cm. Mass 2 000 kg. **Habitat:** Prefers permanent water with submerged sandbanks or gently sloping sandy banks where it can rest by day partly submerged. **Habits:** A nocturnal feeder, it rests by day partly submerged in water. Gregarious, it occurs in schools of about 10 or 15, but may congregate in far larger numbers. Territorial, and males defend their territories against intruders. **Food:** A grazer, it prefers to feed on open areas of short green grass.

Phacochoerus aethiopicus
Warthog
Vlakvark

81.1

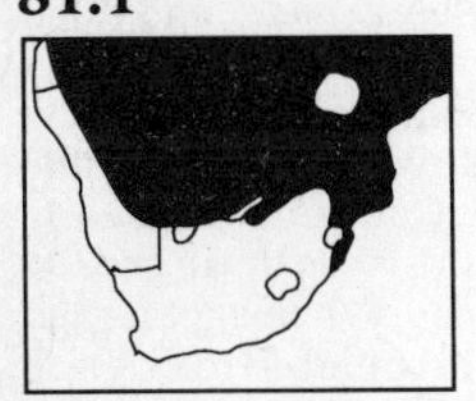

The skin colour, often determined by the colour of the mud in which it wallows, is generally grey. Has characteristic tusks and wart-like protuberances on the face. HB 150 cm. T 40 cm. Ht ♂ 70 cm, ♀ 60 cm. Mass ♂ 80 kg, ♀ 56 kg. **Habitat:** Open ground, grassland, floodplain, vleis and other open areas around waterholes and pans, as well as open woodland and open scrub. **Habits:** Diurnal, it lies up in holes during the night. Wallows in mud, which serves as a means of temperature regulation and protection against biting flies. Sounders consist of an adult male, adult female and her offspring. Maternity groups, bachelor groups and solitary individuals are also found. It grunts to maintain group contact, and snarls and snorts with aggressive display. **Food:** Generally a vegetarian, food includes grasses, especially freshly sprouting grasses, the underground rhizomes of grasses, sedges, herbs, shrubs and wild fruits.

Potamochoerus porcus
Bushpig
Bosvark

81.2

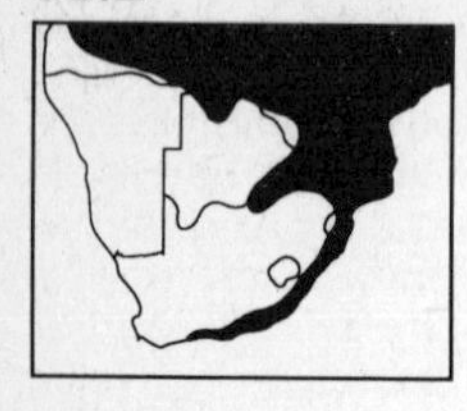

Overall colour varies, but is commonly reddish brown with a dorsal crest of long white hair. Males are slightly larger than females. HB 115 cm. T 40 cm. Ht 55–80 cm. Mass 62 kg. **Habitat:** Forests, thickets, riparian undercover, reedbeds or heavy cover of tall grass. **Habits:** Predominantly nocturnal with some diurnal activity. Gregarious. Sounders consist of a dominant boar and sow, other sows and juveniles. Wallows in mud, probably as a means of temperature regulation and protection against biting insects. When foraging it grunts softly to maintain contact. Lies up in dense cover when sleeping or resting. **Food:** Food includes underground rhizomes of grasses, bulbs, tubers, rhizomes of ferns, corms of nutgrass, wild and cultivated fruits, flowers of lilies, earthworms and the pupae of insects, as well as carrion.

Philantomba monticola
Blue Duiker
Blouduiker

82.1

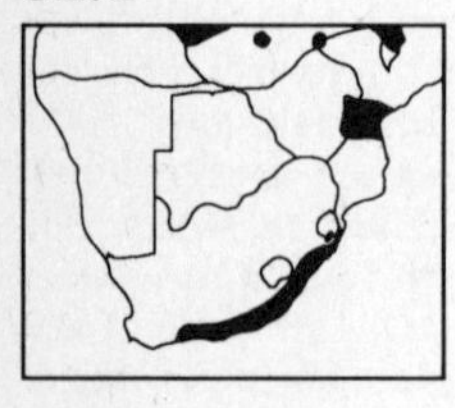

The smallest antelope found in the Subregion. The colour of the upper parts varies considerably, including a dark smokey-brown with a dark bluish sheen, rusty-brown, or dark brown. Has large eyes and slim legs, and both sexes carry tiny horns. HB 58 cm. T 8 cm. Ht 30 cm. Mass ♂ 4 kg, ♀ 4,5 kg. **Habitat:** Confined to forests, thickets or dense coastal bush. **Habits:** Usually solitary, but pairs may associate temporarily, or female with single young. Shy and timid, and at the least sign of danger runs for the cover of thick underbush. Normally silent. **Food:** A browser, it lives on the fine shoots and leaves of low-growing underbush and forbs and fruits fallen from trees. Will also graze on fresh green grass.

Cephalophus natalensis
Red Duiker
Rooiduiker

82.2

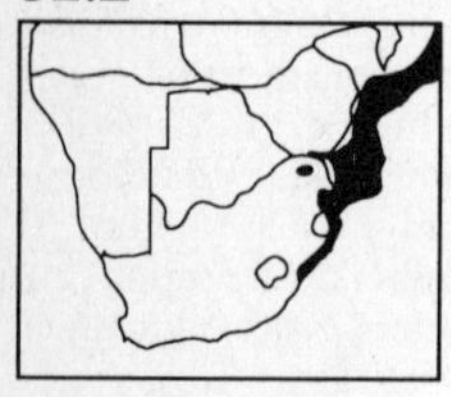

The colour of the upper parts is a deep chestnut-red or an orange-red colour. Both sexes carry short, straight horns. HB 100 cm. T 14 cm. Ht 35–45 cm. Mass 14 kg. **Habitat:** Associated with forests and dense thickets where water is available. **Habits:** Solitary or female with her offspring, or pairs in a loose association. Shy and secretive, it runs for the cover of the thickest bush when disturbed. Mainly nocturnal. **Food:** A browser, it lives on fallen wild fruits, the leaves and fine stems of low-growing shrubs, and dry fallen leaves.

Sylvicapra grimmia
Common Duiker
Gewone Duiker

82.3

The colour varies from a grizzled grey to a yellowish-fawn. Only males carry short straight horns, although occasional horned females are known. Has a characteristic black band on the face from the nose to the forehead. HB 100 cm. T 15 cm. Ht 50 cm. Mass 20 kg. Females larger and heavier than males. **Habitat:** Presence of bush is essential for shelter, shade and food requirements. Will use woodland with ample underbush, but avoids parts that are too open. **Habits:** Solitary, or pairs when females are in oestrus, or female with single young. Mainly active in late afternoon extending well into the night, and in the early morning. Lies up in the shelter of bushes or stands of tall grass at night, and during the hotter hours of the day. When approached it will lie tightly, springing up at one's feet to bound off in a zig-zag course to the nearest cover. **Food:** Almost exclusively a browser, only rarely eating grass.

Neotragus moschatus
Suni
Soenie

82.4

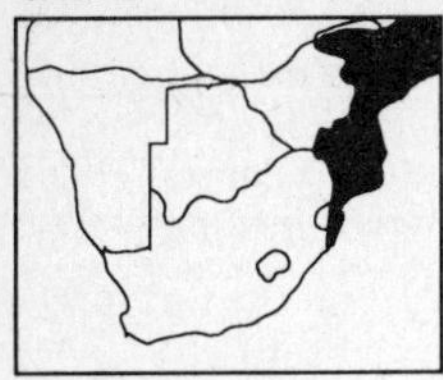

The upper parts are rufous-brown with a slightly speckled appearance; and under parts white. Only males carry horns. HB 60 cm. T 10 cm. Ht 35 cm. Mass ♂ 5 kg, ♀ 5,4 kg. **Habitat:** Dry woodland, with thickets and underbush, riparian scrub or dry scrub along drainage lines. **Habits:** Solitary, pairs or family groups consisting of a male and female with their offspring. Shy and wary. Mainly active in the early morning and late afternoon. During the hottest hours it lies up in dense thickets. **Food:** A browser.

Madoqua kirkii
Damara Dik-dik
Damara Dik-dik

82.5

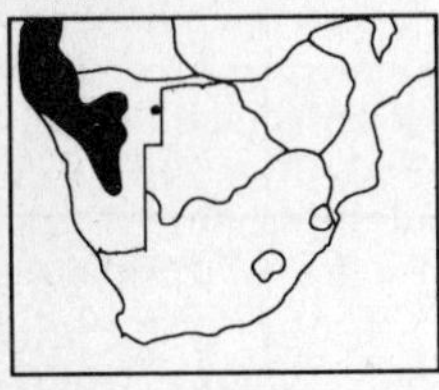

The upper parts of the body are yellowish-grey. Only the males carry small, straight horns. The hairs on the forehead form a distinct crest when erected, and there are white rings around the eyes. Has an elongated proboscis-like, mobile nose. HB 60 cm. T 5 cm. Ht 40 cm. Mass 5 kg. **Habitat:** Dense woodland and thicket with well-developed scrub understorey. **Habits:** Occurs singly, in pairs or in family groups of three. Is shy and, when suddenly disturbed, gives an explosive whistle as it runs for cover. Active at sunrise, in the late afternoon and at dusk, with some activity after dark. During the heat of the day it rests in the shade. **Food:** Predominantly a browser.

Raphicerus sharpei
Sharpe's Grysbok
Sharpe se Grysbok

82.6

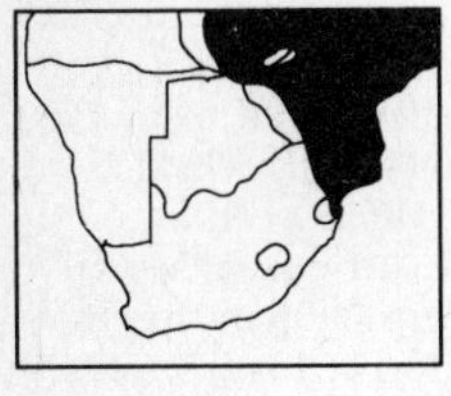

The colour of the body is a rich reddish-brown, sprinkled with white hairs. Only the males carry short horns. HB 70 cm. T 6 cm. Ht 45–50 cm. Mass 7,5 kg. **Habitat:** Areas of low-growing scrub and grass of medium height, avoiding areas of solid stands of high grass. **Habits:** Usually occurs singly, in pairs or female with single offspring. Predominantly nocturnal but also active in the early morning or late afternoon. Lies up in dense cover in the heat of the day. Shy and secretive. **Food:** Predominantly a browser.

Raphicerus melanotis
Grysbok
Grysbok

82.7

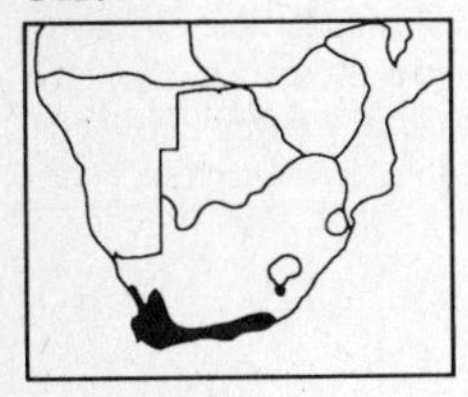

The colour of the upper parts is rufous-brown, sprinkled with white hairs, giving it a grizzled appearance. Only the males have horns. HB 74 cm. T 6 cm. Ht 54 cm. Mass 10 kg. **Habitat:** Thick scrub bush, particularly along lower levels of hills, as well as in broken country or in kloofs, coastal forest or dry succulent veld where there is cover of scrub bush. **Habits:** Mainly solitary, female with young or pairs during the mating period. Nocturnal, it lies up in thick cover during the day. **Food:** Predominantly a grazer.

Raphicerus campestris
Steenbok
Steenbok
82.8

The colour of the upper parts is reddish-brown and the under parts white. Only the males carry horns, although females with horns are occasionally found. The ears are very large. HB 90 cm. T 10 cm. Ht 52 cm. Mass 11 kg. **Habitat:** Associated with open grassland with some cover in the form of stands of tall grass, scattered bushes or scrub. **Habits:** Solitary, except male attending female in oestrus, or female with young. Mainly diurnal. During the hottest time of the day it lies up in cover. Territorial, with established resting places, latrines and preferred feeding places. **Food:** Both browses and grazes.

Ourebia ourebi
Oribi
Oorbietjie
82.9

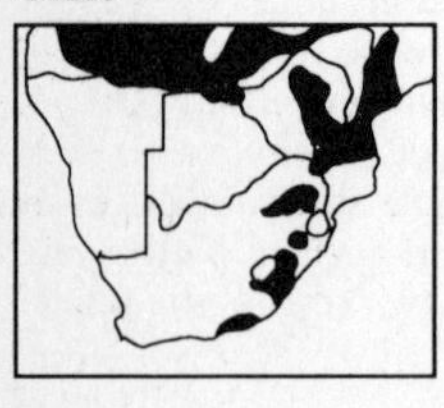

The colour of the upper parts is yellowish-rufous, and the under parts white. The summer coat is shorter and smoother than the winter coat, which is thicker and more shaggy. The tail is black and bushy on the upper surface, and the neck longer and thinner than for Steenbok. Only the males have horns. HB 100 cm. T 10 cm. Ht 60 cm. Mass 14 kg. **Habitat:** Prefers an open habitat, such as open grassland or floodplain, and extensive grassed vleis. **Habits:** Solitary, pairs, or male and one or two females with their offspring. Adult males are territorial. If suddenly disturbed it will give a snorting whistle alarm call as it bounds off stotting with a rocking-horse motion. **Food:** Predominantly a grazer, it will also browse.

Oreotragus oreotragus
Klipspringer
Klipspringer
83

Colour varies from speckled yellowish-brown to greyish-brown and the coat has a coarse texture that blends with the rocks among which it lives. Only the males carry horns. HB 82 cm. T 8 cm. Ht 60 cm. Mass ♂ 10 kg, ♀ 13 kg. **Habitat:** Closely confined to a rocky habitat. Mountainous areas with krantzes, rocky hills or outcrops, rocky koppies and gorges with rocky sides. **Habits:** Generally found in pairs, singly, or in small family groups of a male, female and young. Well adapted to moving around in its rocky habitat, bouncing up steep rock faces or leaping from rock to rock with agility. Often seen standing motionless on a rock pinnacle scanning the surroundings. Active in the early morning and late afternoon, resting up in the shade during the hotter hours of the day. Territorial. **Food:** Predominantly a browser.

Antidorcas marsupialis
Springbok
Springbok
84.1

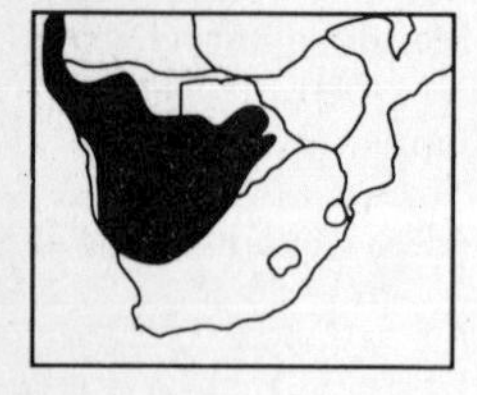

The colour of the back is a bright cinnamon-brown with a distinct broad, dark reddish-brown horizontal band separating the upper parts from the white under parts. Both sexes carry horns. HB 125 cm. T 25 cm. Ht 75 cm. Mass ♂ 41 kg, ♀ 37 kg. **Habitat:** Associated with arid regions and open grassland. **Habits:** Gregarious. Males are territorial. Active in the early morning and late afternoon, with some activity after dark. Solitary males will rest in the open, while herds often seek the shade of low bushes. **Food:** Both browses and grazes.

Aepyceros melampus melampus
Impala
Rooibok

84.2

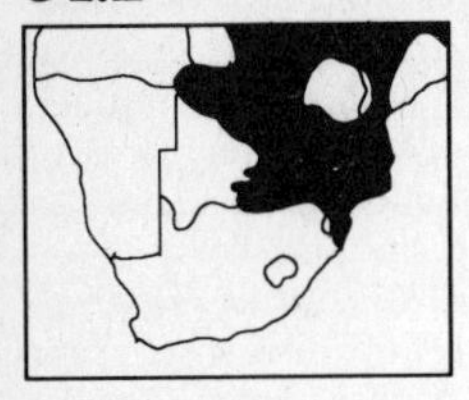

The upper parts are a rich reddish-brown, the flanks are a reddish pale fawn, and the under parts white. On its rump there are distinct vertical black bands. Only the males carry horns. HB 130 cm. T 30 cm. Ht 90 cm. Mass ♂ 55 kg, ♀ 40 kg. **Habitat:** Associated with woodland, preferring light open associations. **Habits:** Gregarious. Social organisation consists of males, which are territorial only during the rut, and bachelor and breeding herds. Predominantly diurnal with some nocturnal activity. During the hotter hours of the day it rests in the shade. **Food:** Browses and grazes.

Aepyceros melampus petersi
Black-faced Impala
Swartneusrooibok

84.3

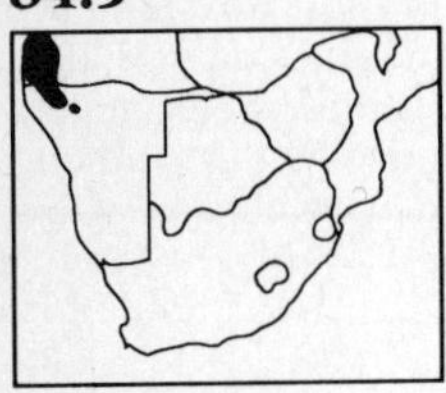

Darker in colour than the Impala. Has a distinct purplish-black band from the nostrils to just in front of the eyes, which continues on top of the head as a thinner band. Mass ♂ 63 kg, ♀ 50 kg. **Habitat:** Dense riverine vegetation bordering on vegetation zones of moderate density. **Habits:** Gregarious. Solitary males are also found. At night herds lie up on open terrain. During the heat of the day small herds lie up in thickets. **Food:** During the rainy season it both browses and grazes, and during the dry season it mainly browses.

Pelea capreolus
Grey Rhebok
Vaalribbok

84.4

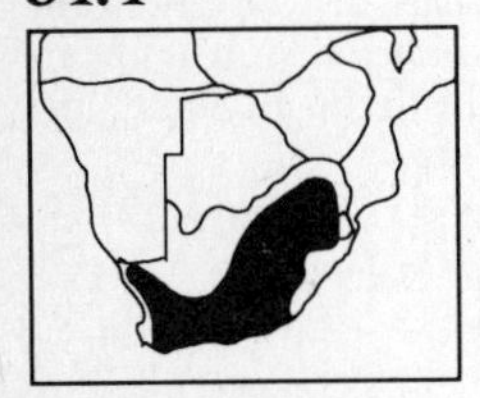

The Grey Rhebok has a long, slender neck and long, narrow pointed ears. The upper parts of the body and flanks are greyish-brown, the under parts are white. The hair is short, thick and woolly. Only the males carry horns. HB 120 cm. T 15 cm. Ht 75 cm. Mass 20 kg. **Habitat:** Associated with rocky hills, rocky mountain slopes and mountain plateaux with good grass cover. **Habits:** Solitary males or family parties consisting of an adult male and several females with young. Males are territorial. Active throughout the day with short periods of resting. Rests during the hotter hours of the day. **Food:** Grazes.

Redunca fulvorufula
Mountain Reedbuck
Rooiribbok

84.5

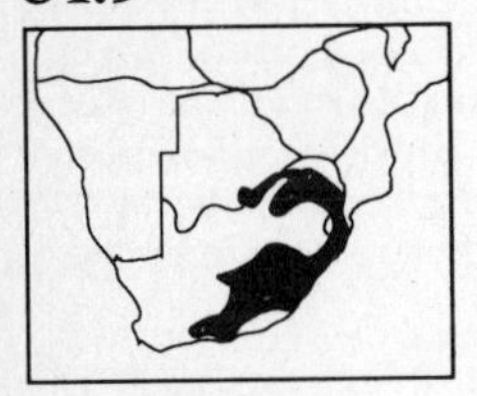

The overall colour is greyish on the upper parts, the under parts white. The coat is soft and woolly with a bushy tail. Only the males carry the short, heavily ridged horns. HB 125 cm. T 25 cm. Ht 75 cm. Mass 30 kg. **Habitat:** Dry, grass-covered, stony slopes of hills and mountains where there is cover in the form of bushes or scattered trees. **Habits:** Social organisation consists of solitary territorial males, solitary non-territorial males or small bachelor groups, and herds of females with young. Most active in the early mornings, late afternoons and at night, resting during the late mornings and early afternoons in the cover of bushes. The alarm call is a shrill whistle. **Food:** Almost exclusively a grazer.

Redunca arundinum
Reedbuck
Rietbok

84.6

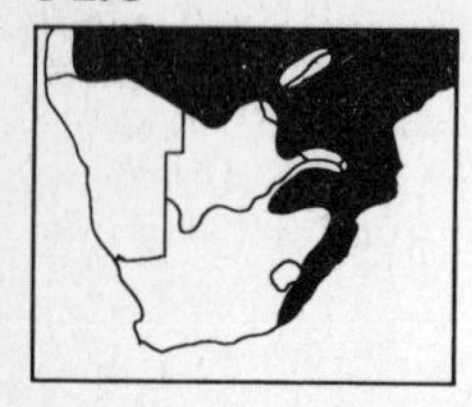

The overall colour of the body varies from brown, tending to grey or buffy-grey, to yellow or buffy-yellow or greyish-brown. The back is usually slightly darker and the under parts are white. Only the males carry horns. HB ♂ 160 cm, ♀ 140 cm. T 25 cm. Ht ♂ 90 cm, ♀ 80 cm. Mass ♂ 80 kg, ♀ 70 kg. **Habitat:** Two essential habitat requirements are cover in the form of tall grass or reedbeds and a water supply: usually vleis with a wet drainage area or grassland adjacent to streams, rivers or other areas of permanent water. **Habits:** Lives in pairs or family parties. Tends to use fixed trails leading to water. Normally rests in the cover of tall grass or reedbeds, but will lie out in the full sun. Whistles when disturbed. Territorial males also whistle to advertise their presence or to establish contact with others. **Food:** Almost exclusively a grazer.

Kobus leche
Red Lechwe
Rooi-lechwe

85.1

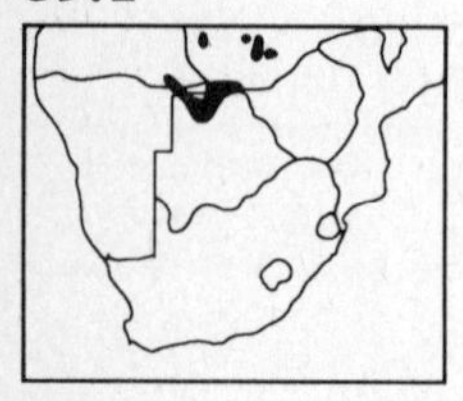

The upper parts of the body and flanks are reddish-yellow and the under parts white. Only the males carry horns. HB 160 cm. T 35 cm. Ht 100 cm. Mass ♂ 103 kg, ♀ 80 kg. **Habits:** Gregarious. Solitary adult males also occur. Active before sunrise and in the early morning, and again in the late afternoon and for a time after sunset, but some may be active throughout the day. Relatively slow on dry land, but once in the water it moves with a plunging gait at a considerable speed. Takes freely to deep water and swims strongly. **Food:** Almost exclusively a grazer.

Kobus vardonii
Puku
Poekoe

85.2

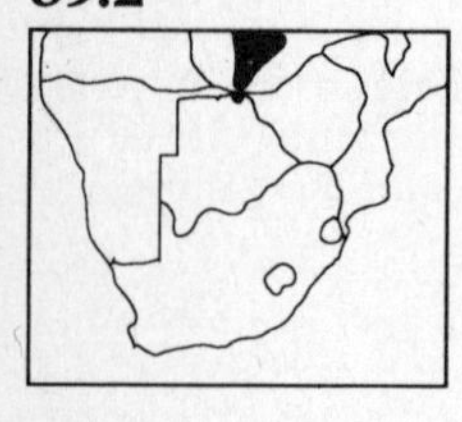

The upper parts of the body are a golden-yellow, the under parts lighter. Only the males carry horns. HB 140 cm. T 30 cm. Ht 80 cm. Mass ♂ 74 kg, ♀ 61 kg. **Habitat:** Associated with grassy areas in the immediate vicinity of water. **Habits:** Gregarious. Adult males are territorial. Crepuscular, with some activity for about an hour after sunset. Produces an alarm whistle up to about five times in succession. **Food:** Predominantly grazes.

Kobus ellipsiprymnus
Waterbuck
Waterbok

85.3

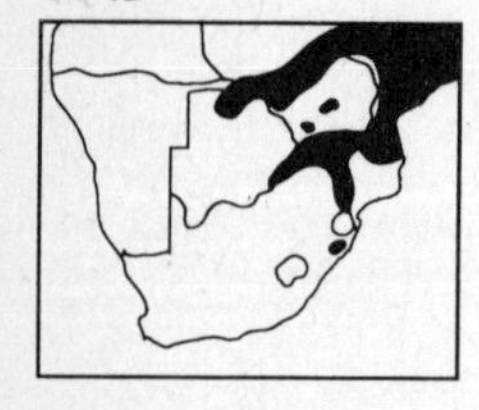

The colour of the upper parts is variable and may be a dark brownish-grey or greyish-brown, grizzled with white and grey hairs. Has a characteristic white ring encircling the rump. Only the males carry horns. HB 210 cm. T 35 cm. Ht 170 cm. Mass 250 kg. **Habitat:** Associated with water. Open areas within reedbeds as in the case of floodplain or with woodland cover. **Habits:** Gregarious. Social organisation consists of territorial males, nursery herds and bachelor herds. Most active in the morning and afternoon to evening, otherwise resting. **Food:** Predominantly a grazer.

Tragelaphus spekei
Sitatunga
Waterkoedoe

86

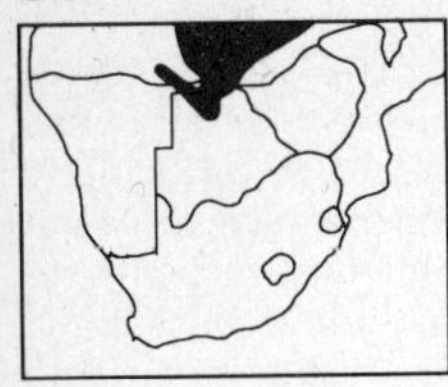

The adult male is a dark drab brown. Its hair is long, coarse and shaggy. The female may be the same colour or redder. Only the male has horns. HB 150 cm. T 30 cm. Ht 90 cm. Mass 115 kg. The females are smaller than the males. **Habitat:** Semi-aquatic, it spends the greater part of its life in dense papyrus and reedbeds in swamp areas in water up to about a metre deep. **Habits:** Active most times of the day, except during the hottest hours when it lies up resting. Also active at night when it moves to the dry fringing woodland. Occurs in small herds consisting of an adult male with several females and juveniles. Solitary males or females also occur. An excellent swimmer, and if disturbed readily takes to deep, open water and swims to safety. Its resting sites are on platforms of broken-down reedstems. **Food:** Aquatic grasses, plants growing in the shallower water, and freshly sprouting tips of reeds.

Tragelaphus scriptus
Bushbuck
Bosbok

87.1

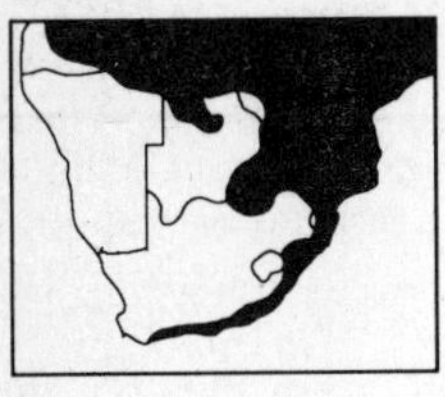

Individuals vary widely in size over their range. The general colour varies from chestnut to dark brown. Has a conspicuous white collar. White spots are found on the cheeks and flanks, and sometimes stripes on the flanks and hind-quarters. Only males carry horns. HB 125 cm. T 20 cm. Ht ♂ 80 cm, ♀ 70 cm. Mass ♂ 40 kg, ♀ 30 kg. **Habitat:** Closely associated with riverine or other types of underbush adjacent to permanent water supplies. **Habits:** Generally solitary. Active at night or in early morning or late evening. During the day it lies up in dense bush. Shy and retiring. When cornered or wounded, the males can be dangerous and under these circumstances have a reputation for aggressiveness. When alarmed it utters a loud, hoarse bark. **Food:** Predominantly a browser.

Tragelaphus angasii
Nyala
Njala

87.2

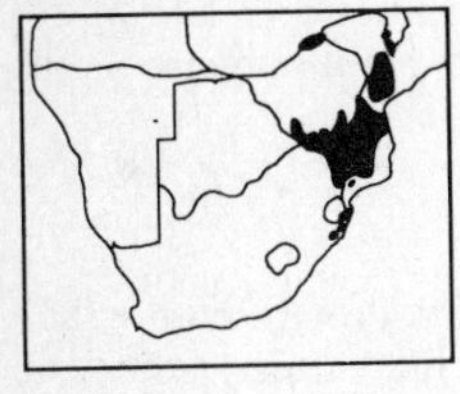

The male is characterised by a greyish shaggy coat with a white crest on the back and long hair on the neck, belly and rump. The sides of the body have 9–14 vertical white stripes, and there are a number of white dots on the body, and a white chevron between the eyes. Only the males carry horns. The females are much smaller, not as hairy, and have a redder coat and more distinct stripes on the sides. Males: HB 165 cm. T 43 cm. Ht 112 cm. Mass 108 kg. Females: HB 140 cm. T 36 cm. Ht 97 cm. Mass 62 kg. **Habitat:** Associated with thickets in dry savanna woodland; closed woodland or more open associations such as found on floodplains; riverine woodland with thickets; and dry forest. **Habits:** Gregarious. Solitary males are also common. Usually silent, its alarm call is a deep bark. Both diurnal and nocturnal. **Food:** Predominantly a browser.

Tragelaphus strepsiceros
Kudu
Koedoe

87.3

The body colour is fawn-grey. Has 6–10 vertical white stripes on the sides. Usually only the males carry horns, but rarely females as well. Males: HB 240 cm. T 43 cm. Ht 140 cm. Mass 230 kg. Females: HB 200 cm. T 42 cm. Ht 125 cm. Mass 160 kg. **Habitat:** Savanna woodland. **Habits:** Gregarious. Adult males may be solitary. Most active in the early morning and late afternoon. During the heat of the day it lies up in woodland or thickets. When alarmed it utters a loud harsh bark. **Food:** Predominantly a browser.

Taurotragus oryx
Eland
Eland

88.1

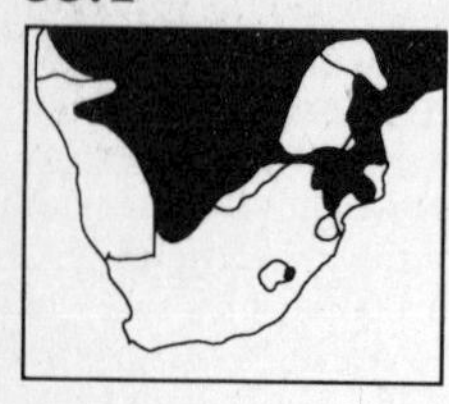

The largest African antelope. The overall colour is a light rufous-fawn, with narrow white stripes down its flanks. Both sexes carry horns, those of the males being much heavier than those of the females. Males: HB 340 cm. T 90 cm. Ht 170 cm. Mass 700 kg. Females: HB 270 cm. T 80 cm. Ht 150 cm. Mass 460 kg. **Habitat:** Occurs in various types of woodland. **Habits:** Gregarious. Active in mornings and afternoons, resting in the shade at noon. Will continue to feed after sunset, particularly during the summer months. Will move on to burnt areas in search of fresh, sprouting grass or to areas where shrubs or trees offer palatable seed pods. Not territorial, and hierarchy is based on age and size. **Food:** Predominantly a browser, but is partial to fresh, sprouting grass after fire.

Syncerus caffer
Buffalo
Buffel

88.2

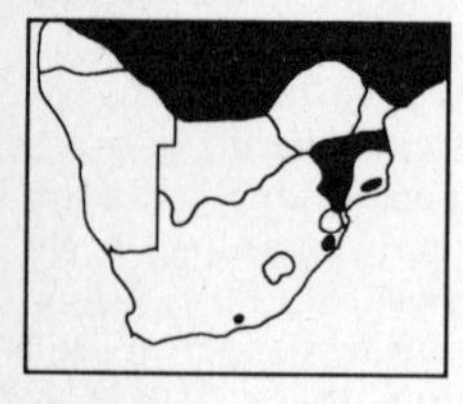

The Buffalo is a very large, heavily built animal, ox-like in general appearance. Old males are black, and females show a tinge of reddish-brown. In old adult males the horns are massive; those of females are always lighter in build. HB 250 cm. T 80 cm. Ht 140 cm. Mass ♂ 800 kg, ♀ 750 kg. **Habitat:** Requirements include a plentiful supply of grass, shade and water. These are found in various types of woodland and open vleis. **Habits:** Gregarious. Small bachelor groups may live independently, and solitary old bulls also occur. Herds have clearly defined home ranges. Most active in evening, night and morning. During the heat of midday it rests in the shade of trees, bushes or reeds. **Food:** Predominantly a grazer.

Oryx gazella
Gemsbok
Gemsbok

89.1

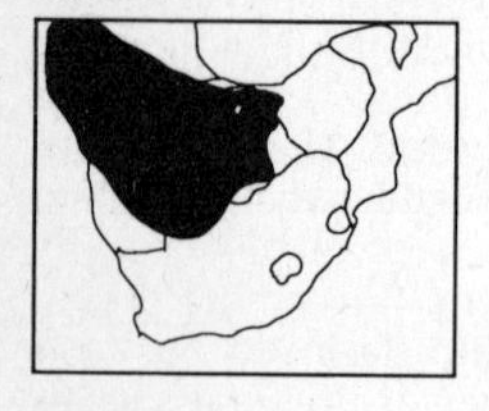

The Gemsbok has very distinct and conspicuous black markings on the body and face. The upper parts and flanks are a pale fawn-grey, the under parts white, with a broad dark-brown band in between. The straight horns are lighter in build in the females than in the males. HB 230 cm. T 90 cm. Ht 120 cm. Mass ♂ 240 kg, ♀ 210 kg. **Habitat:** Open, arid country. Occurs in open grassland, open bush savanna and in light open woodland. **Habits:** Gregarious. Active during the early mornings and late afternoons, lying up during the day. Also active in moonlight. **Food:** Mainly grazing.

Hippotragus niger
Sable
Swartwitpens

89.2

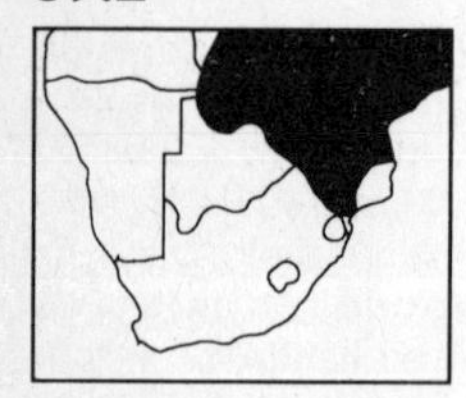

Old adult males are black, females very dark brown, with white under parts. Both sexes carry horns, which are more slender in females. HB 250 cm. T 70 cm. Ht 135 cm. Mass 230 kg. **Habitat:** Savanna woodland. **Habits:** Gregarious. Social organisation consists of territorial bulls, nursery herds and bachelor groups. Bulls are often solitary. Most active in the early morning and late afternoon. Rests in the shade during the middle of the day. **Food:** Predominantly a grazer.

Hippotragus equinus
Roan
Bastergemsbok

89.3

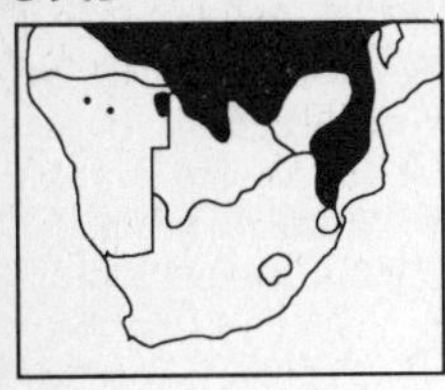

The body colour is greyish-brown. The face is black with contrasting white patch around the nose and mouth, and a white patch on either side of the face. HB 260 cm. T 70 cm. Ht 140 cm. Mass 270 kg. **Habitat:** Confined to lightly wooded savanna with extensive open areas of medium to tall grasses. **Habits:** Gregarious. Social organisation consists of nursery herds, bachelor groups and solitary bulls. Active from sunrise until about mid-morning and late afternoon. During the day it rests up in the shade. **Food:** Predominantly a grazer.

Damaliscus dorcas dorcas
Bontebok
Bontebok

89.4

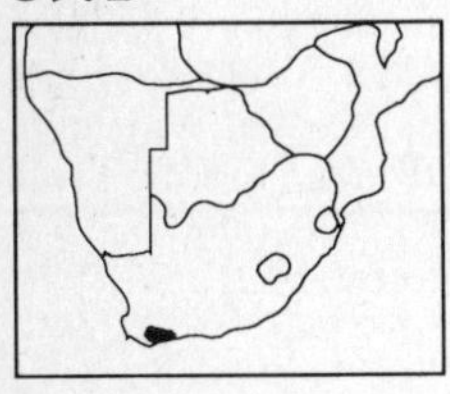

The general body colour is a rich dark brown, the under parts and the rump patches are white. The white face blaze is not divided by a transverse brown band, but is constricted between the eyes. The horns of the female are more slender than those of the male. The pure white patch on the rump of the Bontebok probably distinguishes it best from Blesbok. Ht 90 cm. Mass 61 kg. **Habitat:** Areas of short grass, cover and drinking water are essential. **Habits:** Gregarious, its social organisation consists of territorial males, female herds and bachelor groups. Diurnal, it is mainly active in the early morning and later afternoon. During the hotter hours of the day it rests in thickets where it clusters together. **Food:** Almost exclusively a grazer.

Damaliscus dorcas phillipsi
Blesbok
Blesbok

89.5

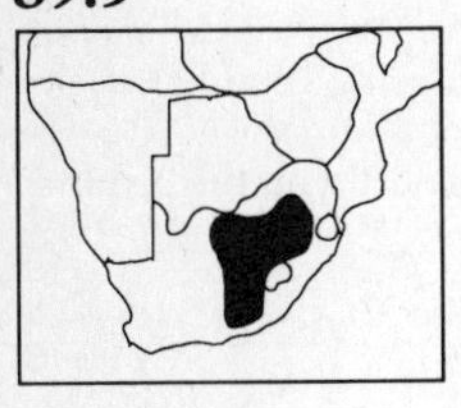

The colour of the body is brown, the under parts are white. The white face blaze is divided just below the eyes by a narrow brown band. Both sexes carry horns. Does not have a white patch on the rump, which distinguishes it from the Bontebok. Ht 95 cm. Mass: ♂ 70 kg, ♀ 61 kg. **Habitat:** Versatile in its requirements, but sweet grasses and water are essential. **Habits:** Gregarious, its social organisation consists of territorial males, female herds and bachelor groups. Diurnal, it is active in the early morning and late evening, and lies up in the shade during the hotter hours of the day. **Food:** Predominantly a grazer.

Damaliscus lunatus
Tsessebe
Tsessebe

89.6

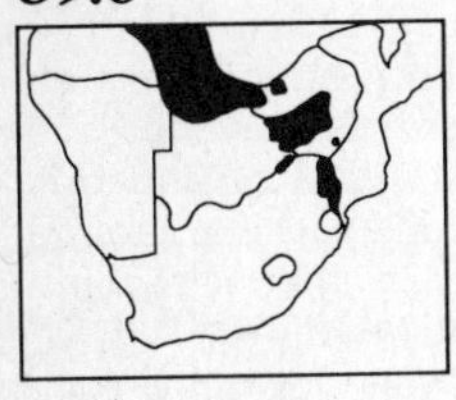

The general colour of the body is dark reddish-brown with a distinct purplish sheen. Both sexes carry horns. HB 170 cm. T 45 cm. Ht 125 cm. Mass ♂ 140 kg, ♀ 126 kg. **Habitat:** Favours the fringes of grassland, at the boundary of woodland. **Habits:** Gregarious. The social organisation consists of territorial males, breeding herds and bachelor groups. Territorial males establish territories which they patrol regularly. Active mainly morning and evening, resting in the shade at midday. The Tsessebe is reputed to be the fastest of any of the antelope occurring in the Subregion. **Food:** Almost exclusively a grazer.

Alcelaphus buselaphus
Red Hartebeest
Rooihartbees
89.7

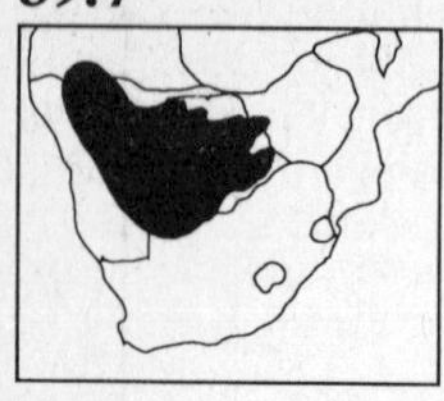

The general colour of the body is reddish brown, but varies and may be yellow-fawn or tawny. Both sexes carry horns, those of the males heavier in build than those of the females. HB 165 cm. T 47 cm. Ht 125 cm. Mass ♂ 150 kg, ♀ 120 kg. **Habitat:** Associated with open country, occurring in grassland, areas of vleis, and semi-desert bush savanna and open woodland. **Habits:** Gregarious. Social organisation consists of breeding herds, bachelor herds and solitary males. Territorial males herd females and defend territories against other males. Most active in the early morning and late afternoon. At midday it may seek shade. **Food:** Predominantly a grazer.

Sigmocerus lichtensteinii
Lichtenstein's Hartebeest
Lichtenstein se Hartbees
89.8

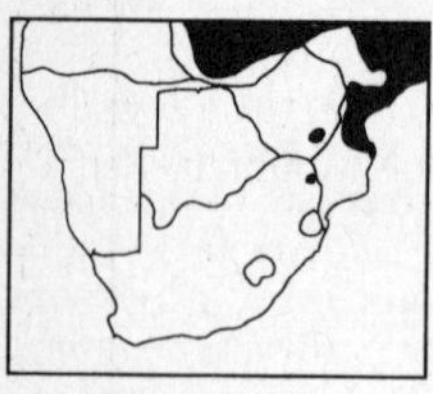

The colour of the body is yellowish-tawny with an indistinct saddle of a more rufous colour. Both sexes carry horns. HB 190 cm. T 48 cm. Ht 125 cm. Mass ♂ 177 kg, ♀ 166 kg. **Habitat:** A savanna species associated with the ecotone of open woodland and vleis or floodplain grassland. **Habits:** Gregarious. It is territorial, and a territorial bull will be accompanied by 8 or 9 females and their offspring. When alarmed the animal vocalises with a 'sneeze-snort' through the nostrils. Has good eyesight, but sense of smell is not so well developed. Most active in the morning and afternoon, resting up in the shade during the heat of midday. May graze at night. **Food:** Almost exclusively a grazer.

Connochaetes gnou
Black Wildebeest
(White-tailed Gnu)
Swartwildebees
89.9

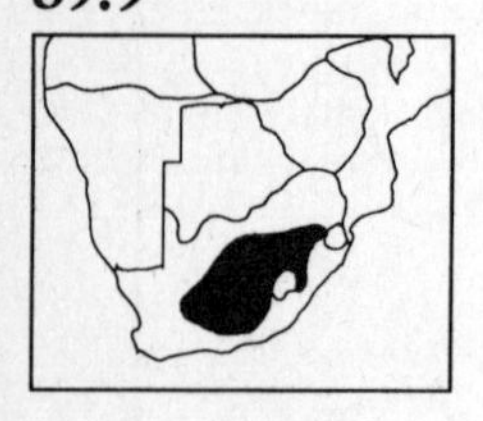

The general colour is buffy-brown. It has a characteristic tail with long, off-white hair reaching to the ground. Both sexes carry horns. HB 220 cm. T 95 cm. Ht 120 cm. Mass ♂ 180 kg, ♀ 160 kg. **Habitat:** Open plains, grasslands and the Karoo country. **Habits:** Gregarious, its social organisation consists of territorial males, female herds and bachelor groups. Active in the early morning and late afternoon, and lies up during the heat of the day. They are also active before dawn and after sunset. On the approach of danger the animal will snort and stamp the ground. **Food:** Predominantly a grazer.

Connochaetes taurinus
Blue Wildebeest
Blouwildebees
89.10

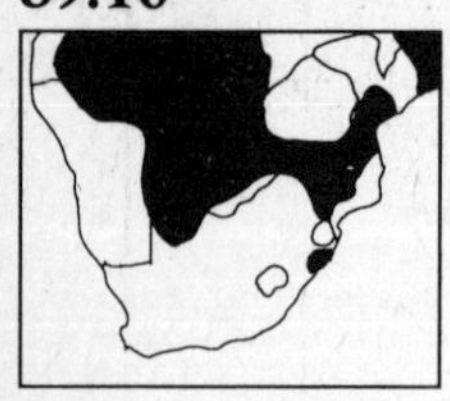

The adult is dark grey in colour. Has a mane of long black hair, with long black hair on the end of its tail. Both sexes carry horns; the horns of the females are lighter in build than those of the males. HB 240 cm. T 100 cm. Ht ♂ 150 cm, ♀ 135 cm. Mass: ♂ 250 kg, ♀ 180 kg. **Habitat:** Associated with savanna woodland, open grassland, floodplain grassland, open bush savanna and light open woodland. **Habits:** Gregarious. Social organisation consists of territorial males, female herds and bachelor groups. While in some parts it is relatively sedentary, in others it is subject to wide movements. In Botswana aggregations of over 100 000 may move to and from the Makgadikgadi Pan on a seasonal basis. When alarmed, adult males will snort. Active in morning and afternoon, it rests in the shade during the hotter hours of the day. **Food:** A grazer.

Giraffa camelopardalis
Giraffe
Kameelperd

91

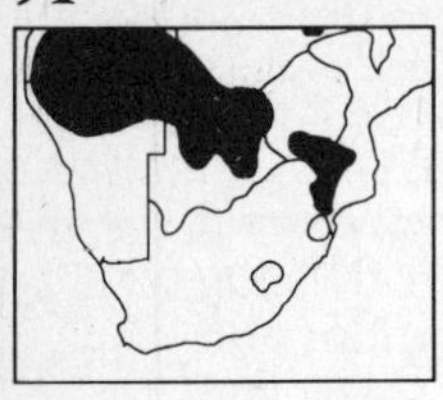

The tallest animal in the world; males are about 5 m high and females about 4,5 m high. Mass: ♂ 1 200 kg, ♀ 830 kg. **Habitat:** A wide variety of dry savanna associations ranging from scrub to woodland. **Habits:** Predominantly diurnal, but will also feed and move at night. Rests during the hottest time of the day. Will put up a fierce defence against Lions. In defence will chop-kick with the front feet or kick with the hind-feet. Males will engage in sparring by swinging their heads at each other, hitting out with their horns. The animal has a loose herd structure consisting mainly of females and young, bachelor herds or mixed herds. Bulls are mainly solitary. **Food:** Predominantly a browser.

Cervus dama
European Fallow Deer
Europese Takbok

92

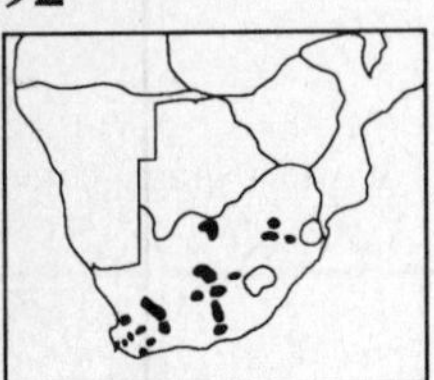

The summer pelage is deep fawn with white spots on the flanks, the winter pelage greyish-fawn and is rough and thick. Many colour varieties are known, including black bodies and others with white bodies. HB 150 cm. T 20 cm. Ht 80–100 cm. Mass up to 110 kg. Females smaller than males. **Habitat:** Catholic in habitat requirements, they have been introduced to a wide variety of vegetational areas. **Habits:** In Europe they are predominantly nocturnal. Their alarm call is a loud bark, which makes them seek safety in cover. Gregarious, in small to large herds. **Food:** Predominantly browsers.

FURTHER READING

This *Concise Guide* is based upon the author's comprehensive and authoritative *Field Guide to the Animal Tracks of Southern Africa* (David Philip, 1990), which contains almost a hundred pages of spoor illustrations and also detailed entries describing the prominent features of the spoor and general information on the animals. In an introductory section, the author explains how to identify and interpret spoor and how to master the basics of tracking.

Louis Liebenberg has also written *The Art of Tracking: The Origin of Science* (David Philip, 1990), which deals with the theory of tracking.

Principal References on Animal Behaviour

Smithers, R. H. N. 1983. *The Mammals of the Southern African Subregion.* Pretoria: University of Pretoria

Stuart, C. and J. 1988. *Field Guide to the Mammals of Southern Africa.* Cape Town: Struik Publishers

Maclean, G. L. 1985. *Roberts' Birds of Southern Africa.* Cape Town: John Voelcker Bird Book Fund

Branch, B. 1988. *Field Guide to the Snakes and Other Reptiles of Southern Africa.* Cape Town: Struik Publishers

Potgieter, D. J., Du Plessis, P. C. and Skaife, S. H. (eds). 1971. *Animal Life in Southern Africa.* Cape Town: Nasou Limited

Skaife, S. H. 1979. *African Insect Life,* new edition revised by John Ledger. Cape Town: Struik Publishers